Successful Simulation

Successful Simulation

A Practical Approach to Simulation Projects

Stewart Robinson
Aston Business School

McGRAW-HILL BOOK COMPANY

London · New York · St Louis · San Francisco · Auckland · Bogotá
Caracas · Lisbon · Madrid · Mexico · Milan · Montreal · New Delhi
Panama · Paris · San Juan · São Paulo · Singapore · Sydney
Tokyo · Toronto

Published by
McGRAW-HILL Book Company Europe
Shoppenhangers Road, Maidenhead, Berkshire, SL6 2QL, England
Telephone 0628 23432
Fax 0628 770224

British Library Cataloguing in Publication Data
Robinson, Stewart
  Successful Simulation: Practical Approach to Simulation Projects
  I. Title
  658.40352

  ISBN 0-07-707622-2

Library of Congress Cataloging-in-Publication Data
Robinson, Stewart
  Successful simulation : a practical approach to simulation
projects / Stewart Robinson.
    p.   cm.
  Includes bibliographical references and index.
  ISBN 0-07-707622-2
  1. Production management—Simulation methods.   2. Simulation
methods.   I. Title.
  TS155.R5944   1994
  003—dc20   94-12966
                                                              CIP

12345 BL 97654

Typeset by Datix International Limited, Bungay, Suffolk
and printed and bound in Great Britain by Biddles Limited, Guildford and King's Lynn

To Jane,
for all her love, friendship and support

and Naomi,
my best 'little' friend

Contents

Preface

In 1986 I started in my first role as a simulation consultant. Some time prior to this I had obtained a degree in Operational Research which included a short course about simulation. However, I was anything but fully prepared to build simulation models, perform experiments and interpret the results. In fact, it would be fair to say that my knowledge was somewhat lacking and without reasonable supervision I would have been a danger to my company, my clients and the good name of simulation.

Having received some initial training I set about work on two or three existing simulation models, making basic changes to the logic before performing experiments. Then, after only two months, I was set loose on some poor unsuspecting client. Doing my best to look confident and professional, I attended an initial meeting to discuss a potential project, after which a proposal and quotation were issued. The proposal was accepted and I soon started work on the model. The project was completed on time and the initial results provided a useful insight into the operation of the facility—so useful that the client asked for more.

To my amazement, not only had I survived this experience, but the project could even have been described as a success. Much of this should be put down to the close supervision I received, a luxury many simulation modellers simply do not have.

Since that early experience I have worked on a large number of simulation projects in both the manufacturing and service sectors, performing the work, managing studies, training modellers and supporting users. The work has been carried out for major national and multi-national companies throughout the world. What I have learned is that if a project is to be successful there are many issues to be addressed, for example: deciding what to include in the model, dealing with inaccurate data, checking the validity of the model, performing experiments and analysing the results. My experience has taught me much about these issues and the lessons are now summarized in the pages of this book.

Objectives of the book

Over the past decade there has been a significant increase in the availability of simulation software. Also, as packages have become easier to use there has been a shift in the knowledge and skills required to build simulation models. Simulation is no longer confined to the expert analyst working on a mainframe computer at the corporate head office. Engineers and operations managers, the 'owners' of the problem, now work on personal computers, building their own models and experimenting to find their own solutions.

So much has the process of model building been simplified that it is easy to forget there is more to simulation than simply building a model. Knowledge and skills are still required to successfully formulate the problem, check the validity of the model, perform experiments and analyse the results. Unfortunately, simulation packages give little help in these areas, which probably reflects the lack of any clearly defined method for performing these steps.

There are many texts available that expound the statistical theory of simulation; however, these are largely too involved and complex to be of much use to the modern simulation practitioner. What is required is a practical approach to simulation projects enabling the non-expert to produce valid results with a reasonable amount of effort.

The aim of this book is to provide such an approach, enabling existing simulation users to improve their effectiveness, while giving new users the tools they require to perform successful projects. Having read this book the reader will:

- Know the stages required for a simulation project
- Be able to formulate a problem for solution by simulation
- Know how to build valid and credible models
- Be able to perform simulation experiments, analyse the results and draw conclusions
- Know how to successfully manage a simulation project

The aim throughout is to provide practical 'rules-of-thumb' and to demonstrate that simulation is more than just a science; it is also an art. It is not expected that the reader will immediately, if ever, work to perfection; however, their performance should be improved. The ideas presented are based on my personal experience and have proved a useful framework for performing simulation projects. Although I have applied many of the ideas to a wide range of situations the reader may need to adapt them for some circumstances.

The book should be read by those who are actually performing simulation projects. It will also provide a useful insight for those who, although not directly involved in modelling, are in some way part of the simulation process, for instance a manager or a member of a project team.

Simulation can be applied to a wide range of problems and to many applications. This book concentrates on its use in Operations Management, particularly in the design, planning, operation and control of manufacturing and service systems.

Contents of the book

The particular features of the book are:

1 *Practical content*　The book is aimed specifically at the non-expert simulation modeller. The theme is practical approaches and solutions to simulation issues at each stage in a project.
2 *Step-by-step approach*　The phases of a simulation project are built up step by step, enabling the book to be used as a general text or as a handbook for reference during the course of a project.
3 *Non-package and non-application specific*　The book is neither aimed at a particular package, nor at an individual application. Specific issues relating to application types are illustrated by examples.
4 *Illustrated by case studies*　Two case studies are used throughout to demonstrate the points being made. The first, Panorama Televisions, develops the theme around a manufacturing problem, while the second, Natland Bank, shows how methods can be applied to service applications. It is recommended that each case study is followed since both demonstrate approaches that can be used in either sector. All the data required to construct these models are supplied, enabling the reader to follow the case studies by building them with their own software. Both case studies are introduced at the end of Chapter 4.
5 *Statistical methods*　The emphasis of the book is on practical methods for performing simulation projects. However, in some instances statistical methods are useful for performing analysis. A chapter is dedicated to explaining some of the main techniques that can be used. This chapter may be treated as optional and only for those wishing to obtain an understanding of how statistics can be applied.
6 *Chapter summaries*　At the end of each chapter there is a summary of the main points. This is designed to act as a quick reference while performing a simulation project.
7 *Further reading*　For those wishing to obtain further insight into the modelling process some additional reading is recommended (see 'further reading' at the end of the book).

Chapter 1 introduces the concept of simulation, where it can be applied and its benefits. Before simulation modelling can begin it is important to select the correct project, tools and project team members; these are discussed in Chapter 2. Chapter 3 gives an overview of a simulation project before a detailed description is given in Chapters 4 to 13.

Chapters 4 to 8 discuss the problem definition phase of the project, including setting objectives, deciding what to include in the model, collecting data and providing a specification. Model structuring, model building and checking the validity of a model are described in Chapters 9 and 10. Chapters 11 to 13 outline the experimental phase, methods for analysing results and project completion.

Chapter 14 summarizes some of the main statistical techniques that can be used in simulation.

The appendices provide some useful societies and journals, a summary of the data in the example models and relevant statistical tables.

Stewart Robinson

Acknowledgements

The author would like to thank the following people who have given their time towards the development of this book:

Carey Adams, Relay Management Training
Neil Higton, Paragon Simulation Services
Geoff Hook, AT&T ISTEL, Visual Interactive Systems
Mike Pidd, The Management School, University of Lancaster

Chapter 1 and Sec. 2.2, stage 1, are based on the papers:

Robinson, S. (1993), The application of computer simulation in manufacturing, *Integrated Manufacturing Systems*, **4** (4), 18–23.

Robinson, S. (1994), An introduction to visual interactive simulation in business, *International Journal of Information Management*, **14** (1), 13–23.

The author would also like to thank Karl Wadsack for his contribution to Sec. 14.5 which is based upon:

Wadsack, K. and Tobias, A. (1994), *The Simulation Project Adviser*, University of Birmingham.

The sample statistical tables in Appendix 3 are taken from:

Murdoch, J. and Barnes, J. (1986), *Statistical Tables for Science, Engineering, Management and Business Studies*, 3rd ed., Macmillan Education, London.

Trademarks

AUTOMOD II, AutoSimulations
INSTRATA, Insight Logistics Limited
GENETIK, Insight Logistics Limited
ProModelPC, PROMODEL Corp.
GPSS/PC, Minuteman Software
SIMFACTORY, CACI Products Company
HOCUS, PE-International
TAYLOR II, F&H Simulations Inc.
SIMAN/Cinema, Systems Modeling Corp.
WITNESS, AT&T ISTEL Limited
SLAM II, Pritsker Corp.
XCELL + , The Scientific Press Inc.
SIMSTAT, MC^2 Analysis Systems
UNIFIT II, Averill M. Law & Associates

PART ONE

Background to simulation projects

1

Simulation, benefits and applications

'What is simulation?' I am frequently asked, and it is in fact a surprisingly difficult question to answer. Most people have some concept of the idea but this is far from complete and it is no simple task to explain it further. Whether it is a senior manager, a colleague or a friend in a restaurant it is useful to have an answer ready, and an answer that works. It is the aim of this chapter to provide such an answer.

Before a simulation project begins there will very often be at least one person involved who has never come across simulation. The first stage is to sell the idea. It is important to explain what simulation is, what its benefits are and where it can be used. If not, these less informed parties will feel threatened and much time will be wasted trying to overcome their resulting resistant attitude.

This chapter begins with an overview of simulation and specifically of visual interactive simulation (VIS), the method employed by most modelling software. The benefits of using simulation are then discussed, followed by an overview of where it can be applied. The chapter concludes with some comments on how to communicate this information.

1.1 Simulation

Since the 1950s computer simulation has been used to tackle a range of business problems leading to improvements in efficiency, reduced costs and increased profitability. Simulation studies have been carried out in most business sectors, including manufacturing and service industries as well as in the public sector.

A simulation is a model that mimics reality; well-known examples are flight simulators and business games. There are many types of simulation. Here we concentrate on the methods employed in Operational Research and specifically discrete event simulation.

Discrete event simulation involves the modelling of a system as it progresses through time and is particularly useful for modelling queuing systems. There are many examples of queuing systems:

- Customers waiting at a supermarket check-out
- Engines queuing on an assembly line conveyor
- Telephone calls arriving at an enquiry centre
- Broken-down machines requiring maintenance
- Vehicles waiting at red traffic lights
- Pallets waiting to be placed into and received from a warehouse

Indeed, a wide variety of systems can be regarded as having a queuing structure and so lend themselves particularly well to discrete event simulation.

A major facet of discrete event simulation is its ability to model random events based on standard and non-standard distributions and to predict the complex interactions between these events. For instance, the 'knock-on' effects of a machine breakdown on a production line can be modelled.

Having built a simulation model (normally on a computer), experiments are then performed changing the input parameters and predicting the response (Fig. 1.1). Table 1.1 shows some examples of inputs and responses for simulation models. Experimentation is normally carried out by asking 'what-if' questions and using the model to predict the likely outcome. It is important to recognize that simulation is primarily a decision support tool and does not directly seek optimum solutions.

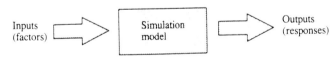

Inputs (factors)   Simulation model   Outputs (responses)

Fig. 1.1   Simulation modelling

1.2   Visual interactive simulation (VIS) and visual interactive modelling (VIM)

VIS software has been available since the late 1970s and represents a significant step forward in simulation modelling. Prior to this, simulation models were largely 'black boxes' which produced a set of results. The model was built and used by an expert model builder while the internal workings of the model remained a mystery to the client (the project sponsor and recipient of the results). Suffice it to say that establishing the credibility of the model and confidence in the results was not a simple task.

Table 1.1 Simulation models: inputs and responses

Model	Inputs (factors)	Outputs (responses)
Supermarket check-out	Customer arrival rate Number of servers Service times	Queue lengths Waiting times Server utilization
Engine assembly	Machine cycle times Machine breakdowns Engine rectification Build schedule Machine changeovers	Throughput Machine efficiency Work-in-progress
Enquiry centre	Call arrival rate Number of operators Operator rosters Call times	Queue lengths Waiting times Number of lost customers Operator utilization
Machine maintenance	Maintenance frequency Repair times Number of repair operators	Machine up-time Operator utilization
Traffic lights	Traffic volume Light change frequency	Average queue length Maximum queue length Average waiting time Maximum waiting time

On-screen animation had been used since the 1960s to portray a running simulation enabling the status of the model to be viewed as it progressed. However, when using VIS the prime motivation is not simply portrayal but interaction with the running simulation. More recently simulation software has been extended to provide interactive capabilities during model building, and hence the term VIM.

VIS/VIM consists of two aspects that set them apart from traditional simulation:

1 *Visual* An animated display of the running model shows the following: elements moving between locations, for example, parts in a factory or customers in a bank; elements changing state represented by colour coding, for example a broken machine is displayed red; and the results in tabular and graphical format such as time series, pie charts and histograms.

2 *Interactive* The simulation is built in small steps, carrying out runs at each stage in order to ensure that the model performs correctly. Once complete, experiments are performed by running the simulation. Part

way through a run the model can be interrupted, changes made to the logic and the run continued. For instance, a lunch time peak in demand at a bank could lead to excessive queues. By interacting with the model, additional tellers can be employed for this peak period and the run continued to obtain the results of this action.

These visual and interactive elements give significant benefits to the model builder and to the client:

- Greater understanding of the model
- Improved communication between all parties involved in a simulation study, removing the credibility gap between model builder and client
- Simpler model validation
- Easier experimentation
- Improved understanding of the results
- Potential for group problem solving
- The use of simulation as a training tool for operations staff

1.3 Why simulate?

The most notable benefits of simulation are:

- Risk reduction
- Greater understanding
- Operating cost reduction
- Lead time reduction
- Faster plant changes
- Capital cost reduction
- Improved customer service

The question must be asked: could these benefits be achieved by other methods such as mathematical modelling and real life experimentation? In some situations these methods would suffice; however, simulation is able to tackle a wider range of issues. There are also a number of managerial reasons for simulation to be used in preference to these alternative techniques.

SIMULATION VERSUS REAL LIFE EXPERIMENTATION

Experimentation can be performed by changing the inputs of the real life system and measuring the resulting change in behaviour. For example, the number of operators at an enquiry centre could be changed until a satisfactory service level is achieved. There are a number of reasons why simulation is preferable to this form of experimentation:

Cost

Real life experimentation can be very costly. It is certainly costly to employ and fire staff at will (if indeed it is possible) or to install additional equipment in order to measure the change in throughput. Having built a simulation model the only additional cost incurred is that of the man-time to change the model and perform the experiments.

Repeatability

Since the exact conditions of a real life experiment are unlikely to be repeated there is only one opportunity to collect the results. Further to this, there is no opportunity to compare the response to alternative inputs under these same conditions. By repeatedly generating the same sequence of events in a simulation model alternative scenarios can be tested under the same conditions. The arrival sequence of vehicles at a set of traffic lights could, for instance, be repeated to test the effect of alternative change frequencies for the lights. Results from different experiments could then be compared and the best scenario chosen.

Control of the time base

When carrying out a real life experiment with the production schedule for a manufacturing plant a month may be required to obtain a result for just one scenario. The same simulation experiment may take a matter of minutes, enabling more scenarios to be examined in a much shorter time period, increasing the possibility of finding a good solution. Sometimes increasing the run time is preferable, turning nanoseconds into seconds, especially when simulating electronic communications and computer networks. Simulation enables such control of the time base.

Legality and safety

Experimentation with new ideas in, say, a toxic chemical plant can be hazardous and even illegal. With simulation, ideas can be tested and once a solution has been found attention can be paid to the safety and legal requirements for implementation.

SIMULATION VERSUS MATHEMATICAL MODELLING

There is an abundance of mathematical models that can be used successfully to represent real world systems, and implementing these models will in most cases be quicker than simulation. Examples of mathematical models

are linear programming, regression analysis and queuing theory. Certainly, simulation should not be used when a simple 'spreadsheet analysis' will suffice. However, there are a number of good reasons why simulation should, in some circumstances, be used in preference to mathematical modelling:

Dynamic and transient effects

Frequently more information is required than just the 'normal' or 'steady-state' behaviour of the system in question. For instance, when modelling a storage facility it is useful to understand the effects of major breakdowns on a high bay crane. It is the ability of simulation to cope with and provide results on these dynamic and transient effects that makes it more effective than mathematical tools.

Non-standard distributions

Data may have been collected on, say, the service time at a supermarket. In queuing theory (as well as other mathematical models) the modeller is restricted to a set of standard service time distributions and so an approximation may have to be made. However, a simulation model can include both standard distributions and distributions based on collected data, enabling the modeller to avoid unnecessary simplification.

Interaction of random events

When a machine breaks down on a production line the stoppage will have significant 'knock-on' effects. Mathematical models cannot easily represent the complex interaction caused by such random events. In general, as the number of random variables increases there is a more than proportional increase in complexity. Simulation is able to handle these complex interactions and predict their effect.

THE MANAGEMENT PERSPECTIVE

It is possible that a problem could be analysed without the use of simulation; however, among the most compelling reasons for its use are the benefits gained by managers. The ability, given by VIS, to view the model running and to interact with it at any stage in the run has significantly added to these benefits.

Simulation fosters creative attitudes

Often, because of the risk of failure, ideas that would give considerable improvement are never tried. However, with simulation, ideas can be tried

in a safe environment and at a low cost; this can only help to encourage innovation and improvement.

Simulation promotes total solutions

There is a tendency for problems to be seen as local issues promoting local solutions. Typically, a build-up of work in progress in one area of a factory is shifted to another department and the problem is 'solved'. A VIS model, showing an overview of the factory, will demonstrate the frailty of local solutions and promote the implementation of total solutions.

Simulation makes people think

When I had completed a simulation project for a major manufacturing company I was asked to discuss the benefits of the work. Having completed this discussion I was told that all the benefits could have been achieved by simply thinking more. However, it is unlikely that the client would have recognized or thought through these problems if the simulation project had not been performed. A major benefit of simulation is that it creates a framework for people to think through specific issues. Indeed, it is possible that 50 per cent of the benefit is achieved before the first experiment has been run by simply gathering information and building the model.

Simulation enables good ideas to be communicated effectively

Many good ideas have been trampled under foot because the benefits could not be demonstrated to a senior manager. The visual aspects of a VIS model prove a powerful tool for communication. It may be that an idea has already been proven but it is deemed necessary to build a simulation model in order to convince senior managers and colleagues of its validity.

1.4 Simulation applications

Simulation is used for a wide variety of applications which are now summarized in eight categories. There is some overlap between these categories and an individual model can be used to tackle more than one of them.

FACILITIES PLANNING

When designing a new facility, simulation is used to ensure that it will perform correctly. By imitating the operation of the facility, bottlenecks are identified, shortages found and solutions sought. For example, a model of a proposed airport terminal could predict bottlenecks in passenger and bag-

gage flows, and might show overcapacity in areas such as check-in and security search. By running experiments with alternative terminal facilities the correct design could be found.

OBTAINING THE BEST USE OF CURRENT FACILITIES

Simulation is not only used in the design of new facilities; a current facility may be performing poorly. Potential solutions may have been identified but these are costly to implement and there is uncertainty as to their success. These solutions could be tested, at a much lower cost, with a simulation of the facility and the best choices identified.

DEVELOPING METHODS OF CONTROL

The design of facilities involves more than just the physical equipment. There are a wide variety of issues concerning working practice and facility control. Experimenting with alternative control logic in a simulation enables the best practice to be identified. For instance, in a manufacturing plant a simulation model could be used to compare the effect of MRP II and KANBAN scheduling.

MATERIALS HANDLING

The flow of materials provides a complex problem in manufacturing and service organizations, for example the movement of vehicles around a manufacturing site or the handling of materials in a warehouse. A model of the material flows and methods of handling would enable congestion points, shortages and weaknesses in control to be identified. Experiments could then be performed to improve the control and flow of materials.

EXAMINING THE LOGISTICS OF CHANGE

The actual process of changing existing facilities will almost inevitably lead to interruptions in the process. In order to minimize these interruptions simulation models are used to examine the logistics of change.

COMPANY MODELLING

Company modelling is particularly useful for simulating operations across more than one location. A high-level model, including only essential details, shows the flows of resources and information between sites. The interaction of the sites is examined and experiments are performed with alternative operating policies.

OPERATIONAL PLANNING

Simulation is used in day-to-day planning and scheduling. A common

application is to test a production schedule by simulating the resulting plant performance. However, simulation has limited application in this area, a subject which is discussed further in Sec. 11.6.

TRAINING OPERATIONS STAFF
Simulation provides a low-risk environment in which supervisors and operators are trained in the operation of a facility. By re-creating specific situations in a model a trainee could attempt alternative actions and discover the consequences. Innovative ideas are encouraged since actions are tested without the risk of incurring significant cost through failure.

1.5   Awareness and education

For any project to be successful it is vital that all the parties involved give their full support. Without this it is possible, if not probable, that the project will fail. Therefore, at the beginning of a simulation study it is important to enlist the support of all those who will in some way be involved. For this to be achieved, managers and colleagues alike need to have a clear understanding of simulation and its benefits.

There is little doubt that simulation is a valuable tool. Risk reduction, improved understanding and lower costs have all been achieved through its use while a wide range of business issues have been tackled. The key issue at this stage is not the success of simulation but developing an awareness of its existence, its relevance and its benefits.

This can be achieved either by educating people at an individual level or by holding group sessions. Remember, the aim is not to explain the intricacies of simulation theory and statistical analysis but to explain how simulation will benefit the organization. A session of about one hour is all that is required. This should describe simulation, why it should be used and on which applications. Where possible, a demonstration of a relevant model is recommended which brings both the idea to life and improves understanding. Above all, ensure that each person is at least giving verbal support before progressing to the next stage.

Summary
What is simulation?

- It mimics reality
- A decision support tool
- Visual interactive simulation (VIS) and visual interactive modelling (VIM)

Why simulate?

- Simulation benefits: risk reduction, greater understanding, operating cost reduction, lead time reduction, faster plant changes, capital cost reduction, improved customer service
- Benefits over real life experimentation: cost, repeatability, control over the time base, legality and safety
- Benefits over mathematical modelling: dynamic and transient effects, non-standard distributions, interaction of random events
- Managerial benefits: fosters creative attitudes, promotes total solutions, makes people think, communicating good ideas

Simulation applications:

- Facilities planning
- Obtaining the best use of current facilities
- Developing methods of control
- Materials handling
- Examining the logistics of change
- Company modelling
- Operational planning
- Training operations staff

Educate managers and colleagues.

2

Selection of projects, tools and teams

A number of choices need to be made before any simulation work can begin. Where there is a choice of projects, which is the best one to perform? If some simulation software needs to be purchased, how can the most suitable be found? Which hardware should be used? What additional software is required?

All these choices will incur some expense. Most simulation software is not cheap; computer hardware can be expensive and projects need both human and physical resources. A justification of the purchase and/or the project is almost certainly required. What factors are involved in such a justification?

The following discussion aims to provide some guidance on how to answer these questions by giving an overview of the issues involved. The expectation is that these will have to be adapted to the needs of the individual organization. For those who require more information some further reading is recommended at the end of the book.

2.1   Project selection

The title of the talk was 'Small is Beautiful!'. An automotive engineer explained how he had saved his company $150 000 in one Friday afternoon. A small simulation model had been built and experiments were performed which proved that the purchase of some equipment was unnecessary. If only most Friday afternoons could be so productive!

'Big is beastly' could best describe this next project. In order to prove how many vehicles were required to move materials around a manufacturing facility, a very large model had been built. After nine months a fully validated monster had been created and experimentation began. A saving of only two or three vehicles was possible which could hardly justify the time it had taken to build the model. To be fair, as the project progressed the scope of the work had become broader and other benefits did accrue.

It would be wrong to suggest that all projects should, or could, be performed on a Friday afternoon. Taking nine months to complete some work may be quite justified when significant savings can be made. However, the principle is that when there is a choice, always select the project which gives the maximum benefit with the minimum amount of effort.

This is especially true when simulation is new to an organization. The first project is often a proving ground and an early success is needed if further studies are to receive support. If a large project is performed first and the results are not available quickly, simulation will soon be seen as time consuming and of little benefit. Building a small but effective model will win the doubters by giving a fast pay-back, enabling credibility to be obtained before moving on to a larger study.

When a large project is the only choice, consider the possibility of separating the work into smaller steps and providing intermediate results. This will overcome the credibility gap by presenting results at regular intervals.

2.2 Selecting simulation software

When selecting simulation software it is important to make the correct choice. If the wrong software is purchased the time required to complete a project may be greatly increased and, at an extreme, the project may not be possible at all. The aim in this section is to give advice on how to go about selecting a simulation package.

A large number of packages are available and at first sight it may seem difficult to make the right choice. However, the process can be simplified by considering two main questions:

1 *Package type*   Which type of package most suits the organization and its structure?
2 *Package selection*   Which specific package best fits the applications to be modelled?

This process can be split further into six stages which are summarized in Fig. 2.1. In stage 1 the type of package is chosen and in stages 2 to 6 the selection of a specific package is made. Each of these stages is now discussed.

STAGE 1: SIMULATION LANGUAGES VERSUS SIMULATORS
Simulation packages fall into two fundamental categories: 'simulation languages' and 'simulators':

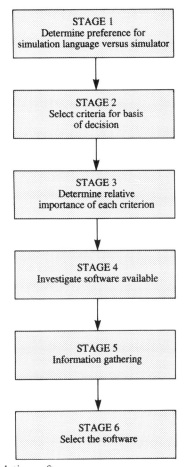

Fig. 2.1   Selecting simulation software

1 *Simulation languages* are general in nature but may have special features for certain types of application such as manufacturing, fast-moving goods and service operations. A model is developed by writing a program using the constructs of the language.

2 *Simulators* are packages that model a specific class of application. They are generally menu driven, requiring a user to input data and basic logic commands while little or no programming skills are required. At one extreme, some simulators are very specific in their application, for example only modelling traffic flows at a set of traffic lights. However, many simulators are very flexible, enabling them to be used on a wide range of applications. Some are able to interface with simulation or other programming languages, providing further flexibility.

Table 2.1   Simulation languages versus simulators

Feature	Simulation language	Simulator
Price	Cheaper	More expensive
Number of applications	More	Less
Modelling flexibility	Higher	Lower
Duration of model build	Longer	Shorter
Time to validate a model	Longer	Shorter
Ease of use	Lower	Higher
Time to obtain modelling skills	Longer	Shorter

Table 2.1 summarizes the main advantages and disadvantages of both types of software.

When purchasing simulation software it is important to consider the suitability of either a simulation language or a simulator in the organization. If projects are to be performed by a central group of experts who are largely dedicated to simulation then a language may be preferred. The higher investment in training is quickly returned while the benefits of greater modelling flexibility are gained. A disadvantage of this approach is that the model builder is one step removed from the problem under investigation. This can lead to difficulties in understanding and communication between the model builder and the client.

In order to remove this gap in understanding, it may be preferred that the clients build their own models. However, it is not cost effective to train a group of non-expert, occasional users in a simulation language. In this case the purchase of a simulator would reduce the training and skills required to perform a project.

Figure 2.2 summarizes the preference for packages in various types of organization. But, there are no fixed rules concerning software selection. Some central organizations have opted for a simulator on the grounds that the results they require can be obtained more quickly while the degradation in flexibility is of little importance. It may also be preferable to purchase more than one package since no software is deemed suitable for all situations.

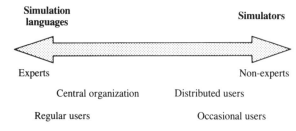

Fig. 2.2   Simulation software preferences

Having decided whether a simulation language or simulator is most suitable, the selection of an individual package can be made.

STAGE 2: SELECT CRITERIA FOR BASIS OF DECISION

At this point it is useful to have a set of criteria in mind as a basis for deciding which package is most suitable. For example, among these could be the ease of model development, the run speed and the hardware platform. In developing these criteria it is helpful if a specific application or set of applications have been identified. It is then possible to determine those criteria which enable these applications to be modelled and so would make the purchase a success.

Table 2.2 provides a list of criteria that could be used. This list is designed to be a guide and is by no means exhaustive. Nor should all these criteria be included when making a decision, but only those relevant to the organization and its applications.

STAGE 3: DETERMINE RELATIVE IMPORTANCE OF EACH CRITERION

The next stage is to make a decision regarding the relative importance of each criterion. Is the ease of model development more important than the run speed? Some form of judgement is required and it may be helpful to assign scores to each criterion on a scale of, say, one to ten. An indication of the suitability of a package can then be obtained by considering how closely it fits, firstly, the important and, secondly, the less important criteria.

STAGE 4: INVESTIGATE SOFTWARE AVAILABLE

Having decided the basis for the decision, it is time to gather information on individual software packages, firstly by exploring which packages are available. A list of some simulation languages and simulators is provided in Table 2.3. This list is neither exhaustive nor is it likely to be up to date since established packages are regularly upgraded and new packages become available. Current information may be obtained by contacting simulation societies, academics and industrial colleagues, or by attending simulation conferences; also a number of journals frequently advertise simulation software. A list of some relevant societies and journals is provided in Appendix 1.

STAGE 5: INFORMATION GATHERING

A short-list of packages can be drawn up by performing some initial enquiries and comparing the findings against a few key criteria. For example, check the application area, the visual features and the price. Most importantly, establish whether the package is a language or a

Table 2.2 Criteria for basis of decision

1 *Suitability for purpose*
- Do the features match the industry/application?

2 *Hardware/software*
- What is the range of hardware platforms?
- Are models compatible between hardware platforms?
- Portability of the software
- Which operating system is required?

3 *Model building*
- Ease of model development
- What debugging aids are available (trace, syntax check)?
- What is the maximum model size?
- Does a model require compilation?
- What is the compilation speed?
- Can data be input from other software?
- Is there a help facility?
- Can a model be built in small steps?
- What model documentation features are available?

4 *Model runs*
- What is the expected run speed?
- Can the run speed be altered?
- What is the interactive capability?
- Is an experimental framework/organizer provided?

5 *Reporting features*
- What standard reports are given?
- Graphical reports—time series, histogram, Gantt charts
- Are graphical reports updated dynamically during a run?
- Is results analysis available?
- Can reports be output to other software?
- Can reports be accessed part-way through a run?
- Can reports be printed directly from the software?

6 *Visual features*
- Is display concurrent with a run or is it a playback feature?

- What resolution of graphics is offered?
- Are icon libraries available?
- Can user icons be drawn?
- How smooth is the movement of icons?
- How quickly can a display be developed?
- Is panning/zooming available?
- Can screen displays be printed?

7 *Statistical features*
- Which standard distributions are available?
- Are distribution fitting procedures included?
- Can empirical distributions be defined?
- How many pseudo-random number streams are available?
- How sound are the pseudo-random number streams?
- Can independent replications be performed?

8 *Support*
- Is there a help desk?
- Can consultancy support be obtained?
- What training is given?
- How frequent are upgrades?
- What is in the next upgrade?
- Is foreign language support available?
- What documentation is offered?
- How good is the documentation?

9 *Confidence*
- Size of the vendor's company
- How long has the package been available?
- Have similar applications been modelled with the package?
- Number of users (in industry sector)
- In which countries is the package used?
- Is much literature available?

10 *Cost*
- Purchase price
- Maintainance fee
- Cost of support
- Cost of training
- How long does it take to learn to use the software?

Table 2.3 Some examples of simulation packages

Simulation languages	Simulators
AUTOMOD II	INSTRATA
GENETIK	ProModelPC
GPSS/PC	SIMFACTORY
HOCUS	TAYLOR II
SIMAN/Cinema	WITNESS
SLAM II	XCELL+

simulator; this will probably enable half the packages to be excluded from the enquiry.

The aim then is to discover how closely each short-listed package matches the decision criteria. Three ways of performing this investigation are now discussed:

Discussion with the vendor

Initially obtain literature and, if possible, a demonstration disk from the vendor. Ask to be given a live demonstration and use this to focus on the features required. A vendor may offer to demonstrate a model relevant to your organization; if not, ask. It may even be possible to get them to build a demonstration model of your own application which only helps to confirm the capabilities of the software.

It is probably best to restrict the discussion with commercial staff to commercial issues while using the vendor's support staff to answer more technical questions. Their answers are likely to be more informed, giving a greater chance of collecting the information required.

Some vendors offer their software on a trial basis. Evaluation by this method requires more time and there may be a fee; however, this can prove a valuable approach. Also, if it is possible to perform a full project within the trial period, this provides a low-cost software loan and some useful results.

A word of caution: it is best to avoid discussion about competitive software. Most vendors only have basic information about their competitors which is often fragmented and out of date. In general, restrict the conversation to the package in question.

Interviewing independent sources

The best method of obtaining impartial information on a software package is to interview an established user. Ask the vendor to arrange a site visit, if possible to a user with a similar application. Even better, find a reference site through another source and so avoid the natural bias of the vendor's choice.

Reading the literature

Ask the vendor to provide some written case studies that demonstrate the successful use of their package. Always request the source of these articles; it is best if they are independent. Also, aim to find articles in relevant journals and the trade press (some useful journals are given in Appendix 1). If a paper is of particular interest contact the author and discuss it further.

STAGE 6: SELECT THE SOFTWARE

Having gathered the information and performed some analysis against the decision criteria it should be possible to select and purchase the correct software. In some cases it may be preferable to purchase more than one package since the diversity of applications requires different modelling approaches.

2.3   Selecting hardware

Choosing the hardware proves more of an issue with simulation than with most other types of software. There is more to consider than just the hardware specification supplied by the software vendor.

Speed is of paramount importance if models of any size are to be built. Performing experiments is time consuming and doubling the speed of a computer halves the time required. The general rule is: the faster the better. Plenty of memory is helpful too.

Building a simulation model is normally a full-time task and so it is best to have sole use of a computer. Trying to share a machine is not effective and only leads to frustration. Aim to obtain hardware that is dedicated to simulation use only.

Finally, the issue of portability is important. If it is likely that the simulation models need to be demonstrated and run at various locations, the model must be portable. Some software packages can be installed and de-installed easily, but with others this is more difficult. If the software is not portable then the hardware needs to be.

2.4   Additional software

With some simulation packages it is either necessary or just beneficial to purchase additional supporting software. For example:

• Spreadsheets

- Databases
- Text editors
- Statistical analysis packages
- Graphics packages

The choice will often be restricted by corporate policy. If more freedom is given, then it is worth discussing the options with the simulation software vendor. Find out which packages their support staff use.

Some packages are available specifically for supporting simulation modelling, for example UniFit II for distribution fitting and SIMSTAT for statistical analysis. This software is often demonstrated at simulation conferences and advertised in relevant journals (see Appendix 1).

When purchasing a simulation package, make enquiries as to which additional software is necessary and which is just 'nice to have'.

2.5 Team selection

Successful projects require strong teams and full participation from every individual. Unless this foundation is established, time-scales will falter and milestones will be missed. Every team member must buy in to the process by understanding where the project is going and what is required of them for it to be a success. It is not the intention here to discuss the art of team building and motivation; there are plenty of courses and books for those who need to develop these skills. What is discussed are the people who make up a simulation project team and the roles that they perform.

Figure 2.3 shows seven roles which are typically required in any simulation study.

THE TEAM LEADER

Team leadership is vital if the direction of the project is to be set, the tasks identified and the time-scales kept. The team leader does not necessarily have to be involved in the modelling process; in fact, it may even be a distinct advantage if team leaders are not, enabling them to concentrate their effort on the leadership of the team.

THE CLIENT

The client is the project sponsor and the recipient of the results. Clients have identified a problem and seek a solution through simulation. They could be the board of directors, a manager, an operator or an external fee-paying customer. Many projects have more than one client with each having different opinions concerning the problem and its potential solutions; one role of the team leader is to ensure that a consensus is reached. Each client should be involved from the beginning of the project and at every

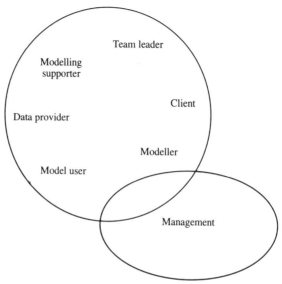

Fig. 2.3   Roles in a simulation project team

stage to ensure that there is consistency in their understanding, input and decisions. However, not every client needs to be involved directly in the team; a client could be represented by one person. In some situations this approach is preferable, for example when the client is the board of directors.

THE MODELLER
The modeller builds and tests the simulation model. Modellers need to have gained expertise in the software package, normally by attending a training course. The modeller can be a full-time simulation expert or work temporarily on simulation for the duration of the project; in many companies engineers and operations staff build their own models. Alternatively, a consultant could be employed to perform the modelling role.

THE MODEL USER
The model may be passed to a user for experimentation. The user is close to the problem and so well placed to seek a solution but users do not need to have the necessary skills to build their own models. However, they do need to be competent in performing experiments and analysing the results. In these circumstances it is important that the model is user friendly and that all the experimental parameters can be changed easily by the model user. This role is particularly necessary when the model is being used in operational planning or for training operations staff (Sec. 1.4).

THE DATA PROVIDER

Simulation requires a great deal of data including layout, timings, flows, control logic and constraints. The data provider is an expert in the system being modelled and either has direct access to the data or knows where it can be found. In more complex studies more than one data provider may be required.

THE MODELLING SUPPORTER

In some instances a third party modelling expert is asked to join the simulation team. This person supports the team in such aspects as model building, model testing and experimentation. Modelling support can be sought from the software vendor, consultants or in-house expertise.

THE MANAGEMENT

Each member of the team has a manager who naturally has a vested interest in the team's performance. From time to time the managers will interact with the process even if they are not directly involved in the team or the project itself. They may require reviews of progress, demonstrations of the model and almost certainly a summary of the conclusions and recommendations. Throughout, it is important to maintain credibility for the work by involving managers in the process.

CO-ORDINATING THE TEAM

On large projects it may be necessary to have one or more persons performing each role. However, teams of more than ten members are difficult to co-ordinate and in this situation the creation of sub-teams should be considered. For example, a group of five clients could form a sub-team and have one representative on the main project team. Another approach is for members of the team to represent certain roles which are not involved directly; for example, the modeller could provide an interface to non-participating data providers.

On smaller projects the team could consist of only two or three members. In such a team an individual performs more than one task; for instance, the modeller is also the model user and the team leader. In some situations there is no need for a team since an individual is able to assume every role.

Whatever the size of the team it is important that each member is identified clearly and participates from the start. Without this, a great deal of time can be wasted bringing new team members up to speed part-way through a project. Selection of the individuals to fulfil each role depends very much on the organization and the skills of the staff employed. Where skills are lacking, ensure that the necessary training is received so that each team member can fulfil their role.

Having selected the team it is useful to set up regular project meetings, probably on a weekly basis. It is easier to cancel a meeting that is not required than to arrange one unexpectedly. The formality of the meetings depends on the working environment and the size of the team. However, it is recommended that minutes are kept since this provides an audit trail of decisions made as the project progresses.

Effort is required to build a project team. It may take some time to find the correct people, to educate them in simulation and its benefits, and to ensure that they all have the necessary skills. However, time put in up-front is not time wasted, and far greater success is achieved when a strong team exists from the start.

2.6 Aspects of justification

The discussion so far has centred on the selection of the project, the team and the tools. Each of these will incur some expense which needs to be justified before a commitment to the work is made. Simulation software often represents an initial outlay of $10 000 to $40 000, and then there is hardware, training and maintenance to consider. The time required by the project team must be taken into account along with the added costs if external consultants are to be employed.

There are four main aspects to be considered in a justification for simulation work:

- Applicability of the technique
- The benefits
- The costs
- Alternative modelling methods

Each of these is now discussed.

APPLICABILITY OF THE TECHNIQUE

Firstly, simulation needs to be established as a suitable method for solving the problem to be addressed. It is unlikely that an issue relating to employees' pay could be resolved with a simulation. On the other hand, simulating a proposed manufacturing facility is probably the best method of ensuring that target throughput is reached. What attributes of a problem make it suitable for simulation?

In Chapter 1 a number of reasons for using simulation are discussed. The most important of these are the ability to:

- Model dynamic and transient effects
- Include non-standard distributions

- Model the complex interaction of random events
- Repeat the conditions of an experiment
- Control the time base
- Safely and legally experiment with new ideas

If at least some of these attributes are present then it is highly likely that simulation is a suitable tool.

THE BENEFITS
The benefits of simulation are discussed in Chapter 1. These are:

- Risk reduction
- Greater understanding
- Operating cost reduction
- Lead time reduction
- Faster plant changes
- Capital cost reduction
- Improved customer service

From the management perspective further benefits are gained:

- Encourages creative attitudes
- Total solutions are identified
- Effective communication
- In-depth analysis of issues

Some of these benefits can be quantified while others cannot. Examples of quantifiable benefits are a potential $100 000 reduction in operating costs and an expected 10 per cent improvement in lead times.

Many benefits cannot be quantified, for instance greater understanding and effective communication. However, these play a vital role and in some instances they are the only benefits to be obtained; an example is a model which shows that a planned facility works without any changes being made. Here the risk of the investment has been reduced and a greater understanding of the facility has probably been gained. Such benefits may well be reason enough for a project to be performed.

The potential benefits need to be identified so they can be weighed against the project costs. For those that can be quantified an estimate of their expected value should be made, most probably in monetary terms. If they cannot be quantified then a list should be created and their worth judged against the project costs.

THE COSTS

The costs of the project can be split into seven areas (typical figures are shown in brackets):

- Software purchase ($10 000–$40 000)
- Software maintenance (5–20 per cent of the purchase price)
- Hardware purchase ($1000–$20 000)
- Hardware maintenance (5–20 per cent of the purchase price)
- Simulation training (free–$10 000)
- Man-time and resources
- Consultancy support ($500–$1500 per day)

At this point it is worth considering whether to ask an external consultant to perform the whole project. The advantages of this approach are that the time-scale is probably shorter and their expertise should improve the quality of the work. The main disadvantages are the cost and the need to educate the consultant in the issues to be addressed. In some cases this route is to be preferred, especially when the purchase of a package is saved or the results are needed quickly. A compromise is to have a consultant build and test the model while the experiments are performed by a model user.

ALTERNATIVE MODELLING METHODS

Finally, as part of the justification, other methods of solving the problem should be considered. Can a mathematical model be built? Is real life experimentation viable? If these other methods suffice then the costs and benefits must be weighed against the simulation approach. Much of this is discussed in Chapter 1.

Conclusion

This chapter has discussed the choice of projects, hardware and software selection, the members of a project team and the requirements for a justification. Now it is time to consider the format of the project itself.

Summary

Project selection

- Choose the project that gives greatest benefit with least effort
- Select smaller projects whenever possible
- Aim to split larger projects into smaller steps

Selecting simulation software

- Simulation languages versus simulators
- Select criteria for basis of decision
- Determine relative importance of each criterion
- Investigate software available
- Information gathering: discussion with the vendor, interviewing independent users, reading the literature
- Select the software

Selecting hardware

- The faster the better
- Obtain sole use of the hardware
- Ensure model is portable through software or hardware portability

Additional software

- Spreadsheets, databases, screen editors, statistical analysis packages and graphics packages

Team selection

- Team roles: team leader, client, modeller, model user, data provider, modelling supporter, management
- The size of the team should satisfy the needs of the project
- Ensure each member is involved from the start
- Ensure skills are developed through training
- Set up regular project meetings and keep minutes

Aspects of justification

- Applicability of simulation
- The benefits
- The costs
- Alternative modelling methods

3

Simulation projects: an overview

When an ascent of Everest begins the climbers almost certainly have some form of plan with details of the route they intend to take, places where they will rest, provisions for each day, possible contingencies and a timing plan. This information is communicated clearly both between the climbers and to the team at the base camp. It is known that during the ascent adverse conditions and unexpected events will cause the programme to change, but nevertheless an initial plan is made.

Imagine the situation without such a plan: what appears to be a good route may end at an impassable ridge; the night may be spent uncomfortably balanced on a ledge; and provisions may run out at an isolated location. The climbers would probably never reach the summit and survival would become the key objective.

Knowing where you are heading, having a plan and communicating the details is vital in any project work. Thankfully, with simulation, failure to do so does not threaten life, but it is likely to threaten the success of the project. Having laid the foundation of understanding simulation and selecting the right resources it is time to consider the process of performing the project itself:

- What are the key phases in a simulation study?
- What are the likely time-scales for completion of the work?

Once this is understood, the project team and the relevant managers can be informed so that correct expectations are set. Note that a detailed plan must wait until the project is better understood and a specification is written (Chapter 8). At this stage only a general outline of a typical project can be communicated.

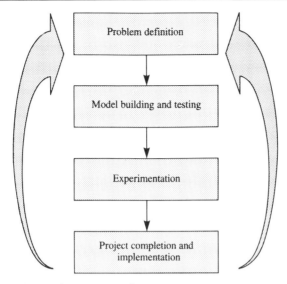

Fig. 3.1   Simulation projects: an overview

3.1   Phases in a simulation project

Figure 3.1 shows that there are four main phases in a simulation project. Despite the fact that these have been shown in a linear fashion, moving from problem definition to model building and so on, the additional arrows aim to demonstrate the iterative nature of the process. A study must always start with a project definition and move towards completion and implementation; however, the movement is not always downwards. For example, experimentation may identify some additional issues which alter the definition of the problem and require further model building before experimentation continues. This iterative process is greatly enhanced by the nature of interactive simulation software, enabling models to be built in stages.

Each phase of the project can be split into smaller steps which follow a similar iterative pattern. An outline of these steps is now given.

PHASE 1: PROBLEM DEFINITION
The initial phase is to understand the problem and to devise an approach for solving it. Figure 3.2 shows the steps which make up this phase.

Identify the problem and set the objectives (Chapter 4)

First the problem is identified; for example, the production of a manufacturing facility is below target or a retail outlet is not achieving the desired level of service. The objectives of the project can then be set, establishing the

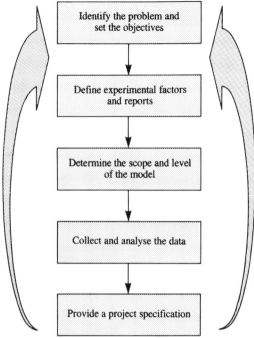

Fig. 3.2 Phase 1: problem definition

issues to be addressed and a measure of what is sought. Examples of object-ives are to increase throughput by 10 per cent, to reduce average waiting time by one minute and to determine the number of vehicles required.

Define experimental factors and reports (Chapter 5)

In order to achieve the objectives particular data are changed and simulation experiments are performed. For example, the number of pallets, the size of buffers and the production schedule could be changed in an attempt to improve the throughput of a manufacturing facility. In this step the relevant data, or experimental factors, are identified.

The success of an experiment is evaluated by one or more measures of performance shown by reports in both tabular and graphical format. These reports aim to show when the objectives have been achieved.

Determine the scope and level of the model (Chapter 6)

Which details should be included in the model and which should be left out? The scope is the breadth of the model, the elements to be represented;

the level is the depth of the model, the details to be included for each element. Only the minimum amount of detail necessary to achieve the project's objectives should be included in the model. This step is also known as conceptual modelling.

Collect and analyse the data (*Chapter 7*)

Having defined the scope and level of the model, the data required are identified. These data are either immediately available or need to be collected. Some data cannot be collected and estimates have to be made. Other data require analysis, for example fitting a distribution to the repair time of a machine. Data collection and analysis may take some time and therefore it is often performed in parallel with the other modelling activities.

Provide a project specification (*Chapter 8*)

A written or sometimes verbal specification ensures that the problem has been properly understood, the objectives are agreed and the modelling approach is correct. Confidence is established before moving to the second phase of the project.

PHASE 2: MODEL BUILDING AND TESTING
The model is built and then tested to ensure that it is a correct representation of the real world. This phase can be broken down into three steps, shown in Fig. 3.3.

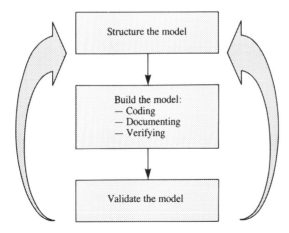

Fig. 3.3   Phase 2: model building and testing

Structure the model (Chapter 9)

Prior to entering the model into the computer the structure is designed, probably on paper. This step ensures that, before the simulation software is used, the best method of modelling is considered. It also provides some useful documentation once the model is built.

Build the model (Chapter 9)

The model is built with the simulation software. This process consists of three distinct activities: coding, entering the model into the computer; documenting, explaining the model structure using the software facilities and other techniques; verifying, ensuring that the code is correct. Each of these activities is performed in small steps, iteratively building and improving the model.

Validate the model (Chapter 10)

Is the model accurate? Are the results realistic? Is it able to meet the project's objectives? Is the model credible? These are all questions that are asked during validation. Validity must be assured before experimentation can begin.

PHASE 3: EXPERIMENTATION

Figure 3.4 shows the experimental phase of the project. Proposed methods for achieving the project's objectives are tested and the results are analysed. During this analysis new ideas are often formed and further experiments are carried out.

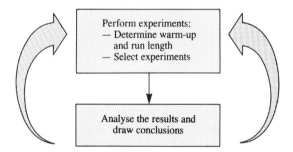

Fig. 3.4   Phase 3: experimentation

Perform experiments (Chapter 11 and 12)

In order to obtain accurate results the model is normally run for a period before any statistics are collected. This warm-up period needs to be determined, as does the length of an experimental run and the number of runs required.

Which experiments should be performed? There may be a large number of potential combinations and time does not allow all of them to be tested. With experimental design a sub-set of these runs is chosen, the results are analysed and conclusions drawn regarding these and other combinations.

Analyse the results and draw conclusions (Chapter 12)

The results of the simulation experiments are analysed. The aim is to check whether the objectives of the project have been met and the extent to which they have been achieved. From this analysis conclusions are drawn and recommendations are made.

PHASE 4: PROJECT COMPLETION AND IMPLEMENTATION
For the project to be complete the results need to be communicated and the recommendations implemented. Also, it is useful to review the successes and failures of the study before performing further simulation work. This final phase is shown in Fig. 3.5.

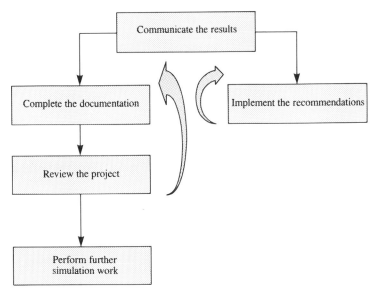

Fig. 3.5   Phase 4: project completion and implementation

Communicate the results (Chapter 13)

The results can be communicated either verbally or with a written report.

Whatever the method, the important thing is that the results are communicated and the conclusions and recommendations are made known.

Implement the recommendations (Chapter 13)

If the recommendations are ignored the project has almost certainly been a waste of time and resource. Ensuring that the ideas generated are actually put into practice is vital for the study to be a success. The implementation is sometimes carried out by a member of the simulation team while on other occasions it is performed by an outsider.

Complete the documentation (Chapter 13)

Various documents are created during the course of the project; these should be completed and made readily available for future use. The model can then be re-used more easily. The documentation also acts as supporting information for a project review.

Review the project (Chapter 13)

It is useful to review the success of the project. The simulation team discusses what was done well, what could have been done better and how it could be done better next time. The purpose of such a discussion is to aim for improvement and to identify specific actions for this to be achieved.

Perform further simulation work (Chapter 13)

Sometimes a new set of issues are identified as a direct result of some simulation work. For example, a model built to consider the scheduling of a manufacturing facility shows a shortfall in throughput. Further studies are then performed to consider these issues in more detail. On other occasions the model is kept up to date with the status of the real facility in order to enable further analysis whenever necessary. Alternatively, a fresh problem could be tackled following the completion of a successful project.

PROJECT REQUIREMENTS

What is discussed above, and throughout the rest of this book, is a 'model' project. The sequence of events and the details of each phase have to be adapted to the needs of the project and the environment in which it is performed. For example, in a small project the specification could be verbal while for a larger project it probably needs to be written.

There is a myth that modern simulation software has removed the need for the 'traditional' project approach. It is true that the ability to build models interactively with a visual display has enhanced and even adapted project methods, but it is not true that it has done away with a structured approach altogether. Recent advances in simulation software have aided the process of model building and to a lesser degree experimentation. However, the problem still needs to be defined, the model still needs to be built and tested, experiments still have to be performed and the results still have to be communicated and implemented. The difference is that each phase can be carried out in an iterative manner, gradually moving from the definition of the problem and moving towards a set of results and recommendations.

A final thought for those who are attempting their first simulation project – obtaining the help of a modelling supporter is almost certainly beneficial. Their greater level of experience is likely to reduce the time required to complete each phase and improve the decisions made. Even if it costs some money to obtain this support, the benefits gained often justify the expense.

3.2   Project time-scales
So far the key phases in a simulation project have been discussed. Our attention now turns to the question of timing.

TOTAL TIME TO COMPLETE THE WORK
It is not possible to give exact advice about the time required to complete a simulation project. Experience shows that they can take anything from a few hours to a few months or even more than a year. A 'typical' project probably requires between one and two months to complete. As a general guide, three main factors can be considered which affect the time-scale most:

1 *Model size*   A simulation of one supermarket check-out with a single queue is about as small as a model can get. At the other extreme, a model of an international airport is likely to have hundreds of service points and hundreds of queues.
2 *Model complexity*   Some models contain only simple logic while others have complex controls for routeing, scheduling and timing, for example, a model of a manufacturing facility demonstrating material scheduling and flows.
3 *Time to experiment*   Sometimes only a few experiments are required to obtain a result. For other models there are many factors and many combinations and a significant number of experiments need to be performed. The speed of the model will also greatly affect experimentation time.

The principle is that as models become larger and more complex and the time to experiment becomes longer so the time-scale increases. Other factors need to be considered too, for example, the amount of data to be collected, whether a simulator or simulation language is to be used, the experience of the modeller and the number of project meetings required.

THE TIME REQUIRED TO COMPLETE EACH PHASE
An estimated percentage breakdown of the project time required to complete each phase is represented in Fig. 3.6. The implementation of the recommendations is not included since this normally continues well after the simulation is complete.

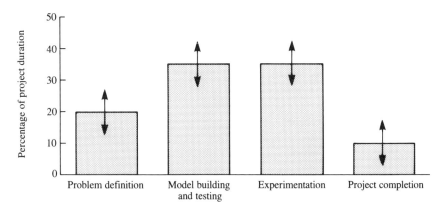

Fig. 3.6   Project time-scales: a percentage breakdown (excluding implementation)

Obviously the actual proportions vary greatly depending on the specific project, especially for experimentation. However, the principle is that the model building and experimental phases take a significant proportion of the time. What is often not taken into account is that problem definition and project completion are also very significant components of the work. These must be allowed for when estimating time-scales.

3.3   Communicate the outline plan
Once the likely phases of the project are understood and a general view of the time-scale has been obtained, it is important to communicate these between the project team and to the relevant managers. Each person is then clear as to where the project is heading and when it is likely to be complete; their expectations are set.

A word of warning: people are anxious to obtain results as early as possible, which in itself is not wrong, but it can lead to a trap. Optimistic time-scales may sound good at the time but when they are not achieved no one gains. Always be realistic, and even a shade pessimistic. If the project keeps to time, no one will complain, and if the project is finished early, the team will receive acclaim for their rapid progress.

During the project regularly review and communicate progress. If the project is falling behind, try to find ways to retrieve the situation. If this is not possible, admit the fault. It is better to warn people that the results will be late than to hope that no one will notice.

Conclusion

The ground work has now been done and the project is ready to start in earnest. Throughout the rest of the book we turn our attention to a detailed discussion on each phase and each step in a simulation project.

Summary

Phases in a simulation project

- Phase 1: problem definition
- Phase 2: model building and testing
- Phase 3: experimentation
- Phase 4: project completion and implementation
- An iterative process

Phase 1: problem definition

- Identify the problem and set the objectives
- Define experimental factors and reports
- Determine the scope and level of the model
- Collect and analyse the data
- Provide a project specification

Phase 2: model building and testing

- Structure the model
- Build the model: coding, documenting, verifying
- Validate the model

Phase 3: experimentation

- Perform experiments: determine warm-up period and run length, select experiments
- Analyse the results and draw conclusions

Phase 4: project completion and implementation

- Communicate the results
- Implement the recommendations
- Complete the documentation
- Review the project
- Perform further simulation work

Project time-scales

- The time required to complete the work depends on: model size, model complexity, time to experiment
- Model building and experimentation are major phases
- Problem definition and project completion take a significant time

Communicate the outline plan

- To the simulation team and to the managers
- Be realistic
- Regularly review progress during the project

PART TWO

Problem definition

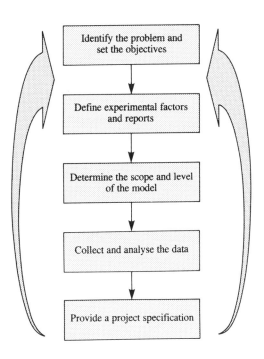

4

Objectives

During a simulation training course I normally run a session on performing projects. I present a problem around which the discussion is based and then ask the question, 'What is the first thing we should do?' It is not unusual for the answer to be, 'collect the data', to which I would reply with the question, 'How do we know which data to collect?' The fact is that, whatever the project, the purpose or objectives must be determined before any such activity begins.

The aim of this chapter is to stress the importance of knowing the objectives in a simulation project and to discuss how they can be set. Firstly, the reasons for having objectives are discussed and then an outline of their structure is given. Some examples demonstrate this structure and the types of objectives that might be expected for different applications. Methods of setting the objectives are then discussed before raising some specific issues relating to general project objectives: the time-scale, run speed, visual and interactive requirements. At the end of the chapter, the case studies that are used throughout the rest of the book are introduced, Panorama Televisions and Natland Bank.

4.1 Why have objectives?

Imagine asking a civil engineering firm to design and build a bridge without ever telling them its purpose. An architectural masterpiece may well be constructed, but if it does not span the right river at the right place, the effort has been wasted. All that needs to be said is that the purpose of the bridge is to connect town A with town B.

This probably seems obvious, and indeed it is. However, it is surprising how often people start a project without ever asking why the work is being done. Perhaps it is our preference for activity and our aversion to thinking that makes us prone to act and then consider the consequences after—a 'shoot first and ask questions later' policy. Or perhaps it is because we are used to making everyday decisions without ever asking why. We eat a meal (in order to stay alive), we go to work (to earn a living) and we go on holiday (to relax). Rarely do people discuss the objectives of the holiday, not because there are none, but because subconsciously they are obvious to all.

The problem is that when more complex projects arise we adopt the same habits: we are so taken with the activity that the purpose is never considered or we assume that the reasons are obvious and they are never communicated to our colleagues. Picture the soccer team which enjoys the game so much that they omit to find out its purpose, namely to score a goal and win. Or possibly the captain knows what to do, but forgets to tell the other players. They may be very skilful and enjoy the game greatly, but if they do not score, their activity is futile.

The principle is that without clear objectives a project is unlikely to achieve success, simply because the work is not directed towards any purpose. Indeed, a successful project has to have objectives which are also communicated and agreed by all. Only then does the activity have direction towards which everyone works so that a useful purpose is achieved.

Therefore, in simulation projects the first step must always be to determine the objectives. Probably more simulation studies fail because this is not done than for any other reason. The objectives set the direction of the work, create the expectation, determine how the model should look, help to identify the experiments, and ultimately judge the success of the project.

4.2   What is an objective?

The objectives state that which should be achieved and the knowledge that should be gained once the project is complete. A useful framework is to consider objectives in terms of three components:

- Achievements
- Measurements
- Constraints

Each of these is now discussed.

ACHIEVEMENTS
Achievements describe the basic aim of the project. For example:

- To increase throughput
- To reduce average waiting time
- To understand the effects of breakdowns
- To determine the number of vehicles

All these contain key actions such as increase, reduce, understand, determine, identify, demonstrate, compare, select and communicate. The action is always performed on something, in this case the throughput, average waiting time, effects of breakdowns and the number of vehicles.

MEASUREMENTS

It is not wholly useful to state that the objective of a project is 'to increase throughput'. By how much is the throughput to increase? Whenever possible a measure of the achievement should be stated. For example:

- To increase throughput **by 10 per cent**
- To reduce average waiting time **by 1 minute**

Ultimately, it is against these measurements that the success of the project will be judged. Has throughput been increased by 10 per cent? Has average waiting time been reduced by one minute? These questions can be answered once the project is complete and used as evidence regarding the success or failure of the work.

In some instances measurements do not apply or it is impossible for them to be quantified. For example, 'to understand the effects of breakdowns' is a valid objective, but it is unlikely that the level of understanding could be quantified. For 'determine the number of vehicles' the measure of achievement is either 'yes' it has been determined or 'no' it has not.

CONSTRAINTS

Consideration should be given to any constraints, or conditions, under which the achievements are to be made. These are normally expressed in terms of money, people, resources or time. For example:

- To increase throughput by 10 per cent **within a capital spend of $100 000**
- To reduce average waiting time by 1 minute **without employing more labour**
- To determine the number of vehicles required **to enable 100 per cent availability of materials**

One advantage of simulation is that experiments can be performed without constraint; therefore, they are not always necessary as part of the objectives.

OUTLINE OF OBJECTIVES

Some examples of typical objectives, for various types of application, are

Table 4.1 Typical simulation project objectives (*measurements are in italics* and **constraints are in bold**)

1 *Facilities planning*
- To reduce capital investment *by 5 per cent*, **while maintaining target throughput**
- To determine warehouse storage requirements **in order to obtain 90 per cent utilization of facilities**
- To improve customer service levels *to 95 per cent* **with the planned facilities**

2 *Obtaining the best use of current facilities*
- To increase throughput *by 10 per cent* **within a capital spend of $100 000**
- To reduce average waiting time *by 1 minute* **without employing more labour**
- To understand the effects of breakdowns

3 *Developing methods of control*
- To compare **three scheduling strategies** *in terms of the service levels achieved*
- To select **one of four methods of organizing labour** *that gives the highest throughput*
- To reduce queueing times in an airport terminal *by 20 per cent* **by only investigating the control of passenger flows**

4 *Materials handling*
- To determine the number of vehicles required **to enable 100 per cent availability of materials**
- To demonstrate the need to change existing handling policies

5 *Examining the logistics of change*
- To demonstrate that plant changes can be made **with negligible interruption to manufacture**
- To identify the level of sub-contracting required during a period of change **so that production is maintained**

6 *Company modelling*
- To reduce intersite movements *by 20 per cent*, **while maintaining profit levels**
- To discover the reasons for material shortages at a site
- To demonstrate the need to relocate a manufacturing plant

7 *Operational planning*
- To reduce order backlogs *by 15 per cent* **by only making improvements to this week's production schedule**
- To identify the number of contract staff required in the coming month **in order to achieve customer service levels**

8 *Training operations staff*
- To highlight the need to repair equipment immediately after a breakdown
- To train the supervisors in methods of selecting jobs **so that throughput targets can be attained**

given in Table 4.1. Note that every objective has an achievement, while not all of them have measurements or constraints.

This outline of objectives is meant as a guide to aid the modeller and not as a framework to be kept at any cost. It is not an absolute formula. There are occasions when objectives cannot be formed in the manner discussed and it would be wrong to try and do so. However, when the objectives are being defined, understanding their structure can be a help.

A final warning: in some instances the purpose of a project is the simulation model itself, with no thought being given to its use. This probably occurs when an organization is so taken with the idea of simulation that having a model becomes an end in its own right. Whatever the reason, this is not a valid objective. Many days are spent creating a model and ensuring that it is correct; then experimentation should begin. The problem is that the objective has been achieved; a model has been built. Even if experimental objectives are found, the model may not be suitable since the relevant details are probably not included. The objectives should always be that which is required of the model, and should never be the model itself.

4.3   Setting the objectives

At the beginning of a project there is a problem that needs to be solved; in some instances little more is known. From a basic understanding of the problem the objectives need to be set, communicated and agreed. What methods are available to ensure that this is done successfully? Some ideas are now discussed.

IDENTIFY THE PROBLEMS

Firstly, the problem needs to be identified. Why is the project being considered? It may be a physical problem, a problem with control or a need to communicate ideas. Whichever it is, the client has a problem that needs to be solved.

This may seem simple, but a client's view of the world is not always correct. The problem with underutilization of a resource may not be lack of control but the result of poor record keeping. The low throughput on a manufacturing facility may not be the effect breakdowns but the product of irregular material deliveries. It is important to try and separate the causes from the effects.

Ask the client what the problem is, both its causes and effects. Listen to the response. When necessary, challenge this opinion and together re-assess the explanation of the causes—even reconsider the effects. Be willing to offer different explanations. Above all, aim to enter a productive and innovative discussion in order to understand the problems more fully.

Of course, if the problems were fully understood then simulation would probably not be necessary since the solutions would already be known. The very purpose of a simulation is to understand the problems and identify solutions. The aim here is not to fully understand the problems but to obtain the best understanding possible. In this way the simulation work can be conducted efficiently and effectively by directing it at the problems in hand.

SET THE OBJECTIVES

Having identified the problems, now it is time to discuss the objectives. What does the client hope to achieve? These should relate to the problems identified:

Ask the client

The direct approach is to ask the client to name the objectives. However, the question, 'What are the objectives?' is not always easy to answer and a better approach may be to ask in a less direct way. For example, 'What do you hope to have achieved by the end of the project?'

Suggest additional objectives

The client does not always realize the potential of a simulation and may not be able to identify certain objectives; always be ready to suggest additional ones. If necessary write them down as a basis for discussion and gain the client's agreement. This proactive approach ensures that the experience of the modeller is added to the knowledge of the client.

RANK THE OBJECTIVES

Some objectives are more important than others. When performing the project it is best to set about the more important objectives first. Find out which are important and which are not, the major and the minor objectives. Then time will not be wasted on a minor problem, such as the number of pallets required, while the real problem is ignored, say to size the warehouse.

DISCOVER ANY FUTURE USES FOR THE MODEL

Discover if there is any intention to use the model for some other purpose in the future. For example, a model of an airport terminal may be adapted for another situation. Such needs should be kept in mind while performing the project. Although the work should not change course to meet these needs, it may be worth a little extra effort to ensure that the model can easily be adapted later.

CHECK FOR CONSISTENCY

Ensure that the objectives are consistent with the problems that have been identified. Also check that the objectives are consistent with each other.

COMMUNICATE AND AGREE THE OBJECTIVES

Having identified the objectives, communicate them to the project team and to the management. Discuss their validity, their scope and their importance. If necessary, change some of the objectives. The aim is to gain agreement before the model is built. The project specification (Chapter 8) is a useful means for doing this.

4.4 General project objectives

Until now the discussion has been restricted to the objectives which directly relate to the identified problems. In addition, there are more general objectives which need to be discussed, namely:

- Time-scale objectives
- Run-speed objectives
- Visual objectives
- Interactive objectives

Each of these should be discussed, and agreed, at the beginning of the project, so the work can aim to meet them.

TIME-SCALE OBJECTIVES

The question of time-scale needs to be discussed with the project team and the management. When are the results required? Is this realistic? If not, can the time-scale (or the objectives of the project) be changed?

When setting the time-scale it is best to identify an actual end date, for instance the beginning of March, and not a time-span, for example in eight weeks time. This prevents confusion over when the project is to be completed.

Time-scales are discussed in more detail in Sec 3.2.

RUN-SPEED OBJECTIVES

Consideration should be given to the speed of a run in order to ensure that the structure of the model facilitates this. If there are a large number of experiments to be performed, then the run speed is an important factor. If the time-scale is tight, a faster model reduces the experimentation time. If experiments are to be performed at group sessions, a run should take a matter of minutes. The aim should be to minimize the run speed while maintaining sufficient detail in the model (Chapter 6).

It may be useful to set a target for the run speed. One method of achieving this would be to provide an estimate by anticipating the number of experiments (Chapter 12) and the number of runs required for each experiment (Chapter 11). However, at this stage an estimate is all that can be given, since exact details of the model structure, the experiments and the length of each run are not yet known.

VISUAL OBJECTIVES

What level of visual display is required? This could range from none at all to a fully animated three-dimensional display. Obviously the latter takes considerably longer to produce and should only be included if it is necessary for success. The decision rests upon the objectives of the project. If the model is primarily a communication tool, then a high degree of graphics is important. The graphics must also be sufficient to aid understanding of the simulation when it is running and provide the information required to test the model (Chapters 9 and 10).

INTERACTIVE OBJECTIVES

The ease of interaction needs to be considered. How should the data be entered, through data files or menu-driven options? Is it necessary to change the data part-way through a run? Should results be available during a run or only at the end of an experiment? These depend on the use of the model and upon the model user. If the model is to be used as a training tool, then a high level of interaction is required. If the model user is an expert, the ease of interaction is less important. At this stage specific interactions are not discussed, but only the level required.

Conclusion

With the objectives set, the project has a firm footing on which to begin. The inputs and the outputs of the model must now be discussed. Which are the main data to be changed during experimentation? What reports should be given to measure the success of an experiment? These are discussed in the next chapter.

Summary

Why have objectives?

- To give direction
- To create expectation
- To determine how the model should look
- To identify the experiments
- To judge the success of the project

What is an objective?

- A statement of that which should be achieved
- Three components: achievements, measurements, constraints

- It is never to build a simulation model

Setting the objectives

- Identify the problems
- Set the objectives: ask the client, suggest additional objectives
- Rank the objectives
- Discover any future uses for the model
- Check for consistency
- Communicate and agree the objectives

General project objectives

- Time-scale objectives
- Run-speed objectives
- Visual objectives
- Interactive objectives

Introduction to the case studies

Throughout the rest of the book two case histories are followed. The first, Panorama Televisions, develops the theme around a manufacturing problem, while the second, Natland Bank, is in the service sector. The aim is to demonstrate how the content of the book can be put into practice. To do this, the case studies have been kept simple, which avoids the simulation issues being clouded by complexity.

It is recommended that both case studies are followed since they demonstrate aspects of a project that could be applied in any sector. As the examples progress it may be useful for the reader to build and experiment with the models using his or her own simulation software. In this way practical experience can be obtained with the ideas that are discussed.

Case study 4.1: Panorama Televisions

Description

Panorama Televisions have been involved in the manufacture of electrical goods since the early days of the radio. They now concentrate on the production of high-quality, premium priced televisions for the international market. There are four televisions in their product range: small, medium, large and flat screen.

Panorama's manufacturing site is shown in Fig. 4.1. Cathode ray tubes (CRT) are assembled in one facility and then transported, by overhead

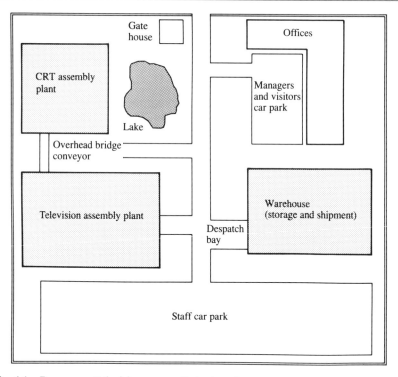

Fig. 4.1 Panorama Televisions: manufacturing site

conveyor, to the television assembly plant. Once the televisions are assembled and fully tested they are taken to the warehouse, stored and then
shipped to the customer. Two forklift trucks transport the televisions from
the assembly plant to the warehouse.

Last year, to meet increased demand, Panorama invested in a new
television assembly plant. Also, after some negotiation with the unions, all
areas of the site moved to continuous working over a five day week.
However, the plant has never achieved its target throughput of 500 units
per day; in fact, throughput is only just over 400 units.

The plant is shown in Fig. 4.2. Plastic moulded boxes are loaded on to a
pallet by an operator at OP10. A production schedule, which is based on
projected demand, determines the type of box to be loaded. At OP20 the
CRT is assembled to the box before the electric coil is added at OP30. The
televisions travel on a conveyor and five manual operators assemble the
electrical equipment, OP40. The television is then tested and any failures
go to the re-work area. Good televisions have the back assembled at OP50
and are unloaded from the line at OP60 by an operator. The empty pallets
are returned by conveyor to OP10; the televisions are stored on a circular

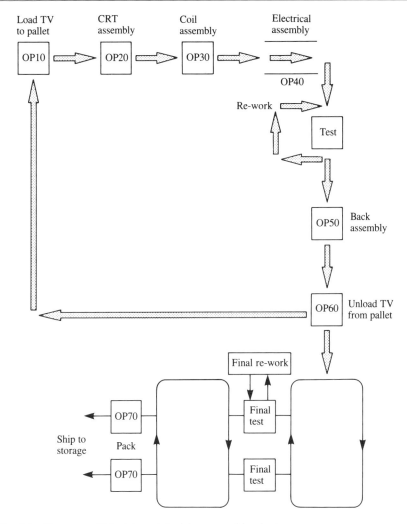

Fig. 4.2 Panorama Televisions: television assembly plant

sling conveyor. A television is taken from the conveyor when a final test booth becomes available. Televisions failing this test are sent for final re-work; televisions passing are stored on another sling conveyor and are packed at OP70. Packed televisions are transported to the warehouse by forklift truck.

The final test and packing area is often short of work and there is enough spare capacity to achieve 500 units per day. The management at Panorama believe that the throughput problem is a result of the number of stoppages on the main assembly line. There are a significant number of breakdowns, and set-ups are required every time there is a change of product in the production schedule. However, there seems little opportunity to improve the efficiency of the machines, nor can the production schedule be changed since it is driven by customer demand. The solution being considered is to increase the buffering between the operations to dampen the effects of stoppages. Design engineers have considered this proposal and believe that, due to physical constraints on space, the buffering could be increased by a maximum of 200 per cent. This will also require further pallets to be bought. In fact, there is some uncertainty as to whether enough pallets are currently being used; increasing the number of pallets may provide a solution without the need for further storage.

Extra storage is expensive so before investing Panorama want to be sure it is necessary. Also, special pallets have to be used at a cost of $1000 each, so it is important to minimize the number required. Target throughput must be achieved, but expenditure should be kept to a minimum.

Objectives
The overall aim is:

- To achieve a target throughput of 500 units per day from the television assembly line

The objectives are:

- To determine whether 500 units per day can be achieved with additional pallets only
- To identify the additional storage and pallets required to achieve 500 units per day

The second objective will only receive attention if 500 units cannot be achieved from the first. The work should be completed in ten working days—by 25 March.

The model is largely required for performing experiments and obtaining results; communication is not a major need. Therefore, the level of visual impact need only enable effective model testing and experimentation.

The interactive capability needs to ensure that basic data changes can be made easily. It would help if results could be accessed at any point during a run.

Case study 4.2: Natland Bank

Description
The Natland Bank is currently experiencing record growth despite a poor economic climate. This is largely due to their policy of offering free loans and massive overdrafts. Consequently, Natland are planning to open a new branch and take advantage of the market potential.

A layout of the branch is shown in Fig. 4.3. The current plan is to have one enquiry desk, five manual tellers and one automatic teller.

Natland are concerned about the facilities they need in order to achieve satisfactory service levels. They are confident, with the exception of one or two managers, that the correct number of manual tellers have been planned and that only one enquiry desk is required. The main concern is over the number of auto tellers. Natland would like to know the customer service level they could expect from one, two and three auto tellers. The bank's policy is that customers should not have to wait for more than three minutes to be served.

Natland have a further choice to make. If more than one auto teller is employed, customers could either wait in individual queues for each teller or in a single queue formed by the use of some plastic barriers. They would like to know which is the best method.

A number of managers are sceptical concerning the need to use a simulation; they have never used it before. They will need convincing that the tool is useful and the results are valid.

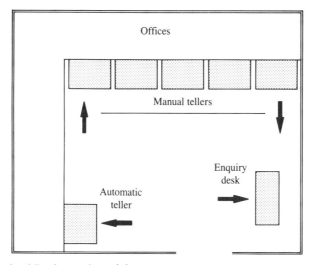

Fig. 4.3 Natland Bank: new branch layout

Objectives

The objectives are:

- To compare the customer service level achieved with one, two and three auto tellers
- To identify which queuing policy gives minimum average waiting times; the choices are a single queue or individual queues for each auto teller
- To convince managers that simulation is a useful tool and the results are valid

If there is time the following objective will also be considered:

- To confirm that the configuration of the other facilities, the manual tellers and the enquiry desk, is correct*

The work should be complete in five working days—by 17 November.

The model should run in less than five minutes, since experiments may be performed at a live demonstration as an aid to communication.

The model requires a high visual impact since it will be used as a communication tool.

The interactive capability should ensure that basic data changes can be made easily, especially the number of auto tellers and the queue discipline. The results need to be accessed at any point during a run.

---

* The fourth objective is added because some doubt has been expressed concerning the facilities other than the auto tellers. This objective is not central to the project and will only be considered if time allows at the end. Alternatively, it could be seen as a separate project, to be performed having completed the investigation of the auto tellers.

5

Experimental factors and reports

The model demonstration was an incredible success—in fact, something of an Oscar winning performance. The audience were amazed at the ease with which data could be changed; they gasped at the three-dimensional iconic display and they nearly stood to applaud the multi-coloured graphical reports. Then suddenly it all went wrong. A manager (who had missed the beginning of the show) asked if the model could predict the effect of employing more service points. The presenter stood and thought, looked to his colleagues for support and then had to admit that it could not. The meeting erupted, the audience left in disgust and the simulation was never seen again.

Even though this is a fictional and exaggerated situation, there is some truth behind it. A simulation model must be able to do what is required of it, otherwise the project cannot be a success. All too often this is not the case and significant re-work is necessary before useful results can be obtained.

Having decided what is required of the work, the objectives, it is time to consider how this can be achieved, the content of the model. In Chapter 1 the basic concepts of simulation are introduced. A model is built to imitate the real world and then experiments are performed, changing the inputs and running the simulation to predict the response. The inputs are experimental factors such as arrival rates, cycle times, schedules, methods of control and failure rates. The responses, or outputs, are shown by reports on, for example, queue lengths, throughput and utilization. It follows that the model must be able to recognize the correct inputs, simulate their effect and report on the resulting response. Otherwise, an experiment may not be possible because the model does not recognize certain inputs or an experiment may not yield vital information because the model does not produce a particular report.

There are three aspects to the model content that need to be considered:

- The experimental factors (inputs)
- The scope and level of the model
- The reports (outputs or responses)

Before deciding on the scope and level of the model, it is important to consider the experimental factors and reports. Only then is it possible to build a model that recognizes the necessary factors and provides the required reports. Therefore, the inputs and outputs are discussed first while the scope and level of the model are covered in Chapter 6.

The aim of this chapter is to show how the experimental factors and reports can be identified, as well as explaining the various ways in which the results can be presented. The chapter begins with a discussion on how to identify the relevant factors and a number of examples are given. Identification of values to be reported and methods of presenting and viewing the reports are then discussed; some examples are also given. The chapter concludes with the Panorama Televisions and Natland Bank case studies.

5.1 Experimental factors

The experimental factors represent the identified methods by which the objectives of the project might be achieved. The objectives describe what should be achieved but no indication is given as to how this might be done. Having set the objectives, the methods of attaining them need to be identified. These methods are then represented by the experimental factors. For example, if the objective is to increase throughput by 10 per cent, then the methods of obtaining this might be to change the cycle times and the buffer sizes. These are also the experimental factors.

There are three decisions to be taken when the experimental factors are being considered:

- Identify the factors
- Determine the range of values that the factors are likely to take
- Decide upon the method of data entry for changing the factor values

Before these decisions are discussed a brief comment is given on their importance in the light of modern interactive simulation software. Although such software enables models to be changed more easily it is still valuable to spend time considering the experimental factors in order to ensure that each can be changed in the simplest possible manner.

IDENTIFY THE FACTORS

Firstly, the various experimental factors should be identified. Brainstorm any potential solutions to the problems which the objectives are aiming to address. Ask the client first, but be ready to offer alternative ideas. Also, the other members of the project team may have thought of different solutions. By creating a list of ways in which the objectives could be met, a list of experimental factors is formed. These could be anything from changing cycle times or employing more labour, to introducing an alternative scheduling policy or designing a new facility.

It may be useful to discuss the relative likelihood of a solution achieving the objectives. Methods that are unlikely to succeed can then be excluded from the model. In this way, time will not be wasted enabling factors to be modelled that have little chance of success.

DETERMINE THE LIKELY RANGE OF VALUES

Having identified the experimental factors it is worth considering the range of values they are likely to take, otherwise known as the factor levels. It is not worth spending days enabling a simulation to model 25 petrol pumps when the forecourt can support only 10. Discuss the maximum and minimum values within which the objectives should be achieved; if anything, be generous in these estimates to prevent the model reaching a premature limit. Some factors may simply have an 'on' or 'off' level, for example the use of just-in-time scheduling. For some experimental factors preliminary analysis of the available data could be used to identify the range of values within which potential solutions are expected to lie.

DECIDE UPON THE METHOD OF DATA ENTRY

The final decision is the method of data entry for the factor values. This can be done in a variety of ways:

- The simulation model code
- Menu-driven options
- Data files
- Third party software such as spreadsheets and databases

The need to perform interactive experiments and change these factors part-way through a run should also be considered.

The choice is driven by the use of the model and the needs of the model user. What experience does the user have? How easy should it be to make the changes? Is the model to be used for interactive experiments? The interactive objectives, discussed at the end of Sec. 4.4, should act as an aid in making these decisions.

Table 5.1 Examples of experimental factors

1 *Facilities planning*
Objective: to improve customer service levels to 95% with the planned facilities

Factors	Levels	Data entry
Number of staff	15 to 30	Data file
Staff roster	Fully flexible	Data file

2 *Obtaining the best use of current facilities*
Objective: to increase throughput by 10% within a capital spend of $100 000

Factors	Levels	Data entry
Cycle times	$\pm$ 10%	Model code
Buffer sizes	$\pm$ 50%	Model code

3 *Developing methods of control*
Objective: to select one of four methods of organizing labour that gives the highest throughput

Factors	Levels	Data entry
Method 1	On/off	Menu driven
Method 2	On/off	Menu driven
Method 3	On/off	Menu driven
Method 4	On/off	Menu driven

4 *Materials handling*
Objective: to determine the number of vehicles required to enable 100% availability of materials

Factors	Levels	Data entry
Number of vehicles	50 to 100	Data file

5 *Examining the logistics of change*
Objective: to demonstrate that plant changes can be made with negligible interruption to manufacture

Factors	Levels	Data entry
Method of change	1*	Model code

6 *Company modelling*
Objective: to reduce intersite movements by 20%, while maintaining profit levels

Factors	Levels	Data entry
Number of vehicles	100 to 200	Menu driven
Size of vehicles	1 to 32 tonnes	Menu driven
Facility location	10 sites	Menu driven

7 *Operational planning*
Objective: to identify the number of contract staff required in the coming month in order to achieve customer service levels

Factors	Levels	Data entry
Number of contract staff	0 to 10	Spreadsheet
Shift patterns	Fully flexible	Spreadsheet
Customer demand	0 to 1000	Database

8 *Training operations staff*
Objective: to train the supervisors in methods of selecting jobs so throughput targets can be attained

Factors	Levels	Data entry
Number of jobs	0 to 10 000	Data file
Type of jobs	Type A to type R	Data file
Sequence of jobs	Any sequence	Menu driven

*Only one factor level is assigned because the objective of the model is to demonstrate that plant changes can be made using this one method only.

Some examples of experimental factors are shown in Table 5.1. These are related to the modelling applications and some of the objectives outlined in Chapter 4, Table 4.1. This is not intended to be an exhaustive list but a demonstration of the range of factors that might be required. Now that the experimental factors, their levels and method of entry have been identified, the model can be structured to ensure that the necessary experiments are possible with a minimum amount of effort (Chapter 6).

A final comment: the experimental factors may well change as the project progresses; it would be unusual if every solution could be thought of at this early stage. During the process of analysing the data and building the model further potential solutions are often uncovered. This does not render invalid the need to identify potential solutions and experimental factors at the beginning of a project. However, it does emphasize the need to follow an iterative approach to the work.

5.2   Identifying the reports

There are three stages in identifying the reports that a model should give:

- Identify the values to be reported
- Determine the method of reporting
- Decide how the reports should be viewed

These are now discussed. In this section the identification of the values is covered, while Secs 5.3 and 5.4 discuss the methods of reporting and viewing respectively. Some examples are given at the end of Sec. 5.4.

IDENTIFICATION OF VALUES TO BE REPORTED

At this stage it is important to recognize that the reports have two basic roles:

- Measuring the extent to which the project's objectives have been achieved
- Pointing to problems that are preventing the objectives from being achieved and so enabling solutions to be identified

The values to be reported should, by relating them to the objectives, provide the information that supports both of these roles.

For example, if the project's objective is to increase throughput by 10 per cent, a report on throughput is required. Results showing the utilization of machines and the size of buffers are also helpful in pointing to any problems. If an experiment is not successful in giving the increased throughput, an analysis of these reports should point to the cause and thus to any potential solutions. For instance, a full buffer might be blocking a machine so increasing the buffer size could provide a solution.

Always ensure that the model calculates these values in the same way as in the real world. Where the simulation package automatically reports a value, ensure that the calculation conforms. Otherwise, at best there will be confusion and at worst a decision taken based on faulty information. If the real world method is not the best, use another one, but be sure that the difference is made clear. On one occasion, having completed a study on a manufacturing line, I gave a presentation to the manager. Although the presentation was well received, a lot of time was wasted trying to understand and explain a discrepancy in the results on component repairs. It turned out that the calculation in the model was at variance with the method used in the real world. If the calculation had been the same, precious time would not have been lost.

Having decided which values need to be reported, the methods of presenting them should be considered.

5.3   Methods of reporting

There are a number of ways to present the reports. It is important that the right methods are chosen; this ensures that the results can be interpreted correctly and also communicated effectively. The aim in this section is to describe various methods of presenting the reports and to give some advice on the advantages and disadvantages of each.

Before discussing specific methods of reporting, there are some general issues to be addressed. Firstly, many of the reports described below should be used in collaboration with each other and not in isolation. For instance, when a histogram is being used, the mean, standard deviation, maximum and minimum should normally be reported too. Secondly, most simulation packages give facilities for producing many of the reports outlined below. In some cases reports are given automatically; in others they may have to be defined by the modeller. Finally, account should always be taken of the way in which the results are presented in the real world. If a pie chart is normally used, it is probably best to provide one in the model to take advantage of their familiarity. If the method used is at variance with the real world, make this clear.

What methods are available for presenting the reports?

TABULAR REPORTS

In this section, the meaning of a tabular report is very general. In other words, a table could be anything from just one figure to a large array of numbers. The format of the table is not covered but only the information that it provides. Six ways of presenting a result are discussed:

- Cumulative total and percentage

- Mean and standard deviation
- Median and quartiles
- Mode
- Minimum and maximum
- Statistical tests

These are now explained.

Cumulative total and percentage

It is useful to know the cumulative total of some values, for example total throughput, total work in progress or total time spent serving customers. On other occasions the total can be expressed as a percentage, for instance the percentage of customers served within five minutes of arrival or the percentage of time a machine is idle. A shortcoming is that no indication of variation is given while modelling variability is one of the major reasons for performing a simulation study.

Mean and standard deviation

The mean average is commonly referred to as just the average. For example, the mean daily throughput is calculated at the end of a simulation run as an indication of the facility's average throughput. The mean is calculated as follows:

$$\text{Mean} = \frac{\text{value}_1 + \text{value}_2 + \cdots + \text{value}_n}{n}$$

where n is the number of values recorded. For example, the mean daily throughput over a ten day run is

$$\frac{\text{Throughput on day 1} + \cdots + \text{throughput on day 10}}{10}$$

In itself, the mean gives no indication of the amount by which the individual values vary. However, this would be useful information. For example, there is a significant difference between a facility that gives a steady throughput and one where the throughput varies greatly, even though the mean may be the same. The data from which the mean is calculated contain this information and can be summarized using the standard deviation:

$$\text{Standard deviation} = \sqrt{\frac{(\text{value}_1 - \text{mean})^2 + \cdots + (\text{value}_n - \text{mean})^2}{n - 1}}$$

For example, the standard deviation of daily throughput over a ten day run is

$$\sqrt{\frac{(\text{Throughput on day 1} - \text{mean daily throughput})^2 + \cdots + (\text{throughput on day 10} - \text{mean daily throughput})^2}{9}}$$

The standard deviation is greater when there is more variation in the values.

The mean and standard deviation are the most commonly used measures of the average and spread. Most simply, they provide a useful means for inspecting and comparing the results from different experiments. However, they take on particular meaning when performing further statistical analysis (Chapter 14).

Median and quartiles

An alternative way of expressing the average is to use the median. This is the most central measurement. If all the values are arranged in order of magnitude then the median is the middle value. For example:

Values:	345	398	503	420	457	234	367
Arranged values:	234	345	367	398	420	457	503

$$\uparrow$$
Median

If there are an even number of observations, the median is calculated as half-way between the two central values.

In the same way that the standard deviation describes the spread of values about the mean, so quartiles express the spread around the median. A quarter of the values lie below the lower quartile while three-quarters of the values lie below the upper quartile. The difference between the upper and the lower quartile is known as the interquartile range. This is a useful measure of the spread.

A frequency chart of daily throughput is shown in Fig. 5.1 from which a cumulative frequency chart has been constructed. The median, lower quartile and upper quartile are all shown.

The median is a useful measure when the values are likely to be significantly skewed. When looking at the average size of a queue, it might be that the queue is often fairly small and just occasionally it is very large.

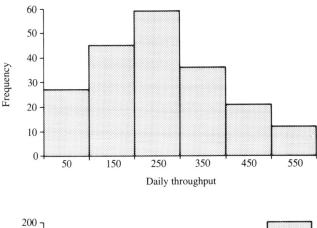

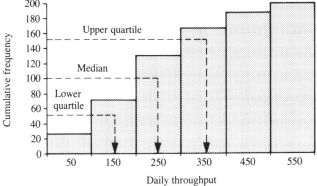

Fig. 5.1   Median and quartiles of daily throughput

The mean would tend to give a lot of weight to the large values while the median would not. Therefore, the median might be the preferred measure in this case. The main disadvantage of the median is that, unlike the mean, there is little in the way of further analysis that can be performed with it.

Mode

The third way of expressing the average is the mode. This is the value that occurs most frequently; therefore, for example, the most common value of daily throughput is the mode. It is most useful when a particular value occurs frequently. However, in general its use is not recommended since no indication of the spread of values is given.

Minimum and maximum

The simplest way of measuring spread is to report the maximum and minimum values obtained. These are very useful when shown in conjunction with other measures of spread such as the standard deviation. However, the possibility of outlying values giving an extreme maximum or minimum means they should not be used in isolation.

Statistical tests

Statistical tests such as confidence intervals can be useful ways of expressing and analysing the results. These are discussed in the chapter on statistics (Chapter 14).

GRAPHICAL REPORTS
There are five main types of graphical report:

- Time series
- Histograms
- Gantt charts
- Pie charts
- Scatter diagrams

In the following discussion each of these is explained along with an outline of when they could be used.

Time series

A time series records the level of some value at regular intervals of time, such as hourly, daily or weekly. They are often used by the press for reporting key economic data such as unemployment figures and inflation rates over the past twelve months. However, they are rarely referred to as time series in this context. Figure 5.2 is an example of their use in a

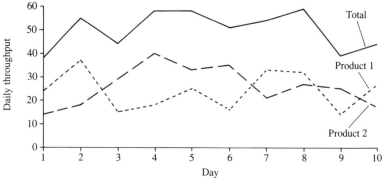

Fig. 5.2 A time series showing daily throughput

simulation showing the daily throughput of two products and the total.

Time series are one of the most useful reports in a simulation, showing, for example, the changes in throughput, work-in-progress and average queue sizes over time. They are especially useful in showing the history of a simulation run, which can in turn identify periods when problems occurred, for instance a period of low throughput. Time series also indicate the variability of a value which is important when modelling random events.

The main disadvantage of using a time series is that only an indication of variability is given; it does not show a value's distribution. Also, valuable information is lost since a time series can only maintain a record of values over a limited window of time, for example 24 hours. This can be overcome by writing the full history to a data file (Sec. 5.4) as the model runs.

Histograms

A histogram records the variation in the level of a value, in other words its distribution. They are also known as frequency charts, bar charts and column charts. Figure 5.3 is an example showing the distribution of waiting times for customers at a supermarket check-out.

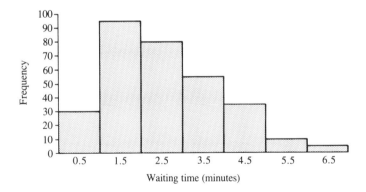

Fig. 5.3   A histogram showing waiting time at a supermarket check-out

Stacked bar charts can be used to show additional information about the values recorded in a histogram. Each bar is split into a number of ranges and either shaded or coloured differently. In Fig. 5.4 the histogram shown in Fig. 5.3 has been shaded to show the waiting times at each individual check-out.

Histograms clearly show the variability in a model. They are often used to complement a time series, showing, for instance, the distribution of daily

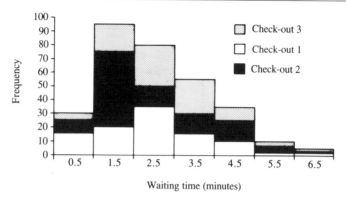

Fig. 5.4 A stacked bar chart showing waiting time at each individual supermarket check-out

throughput. Histograms are also used for recording values that cannot be collected at regular intervals of time, for example the repair times of a machine. A major use is to validate the samples taken from an input distribution, say the interarrival times of customers; the histogram is then compared to the data that were input to the model.

Their main disadvantage is that no account is made of time; the history of events is lost. For this reason, when it is applicable, the use of a time series with a histogram is recommended.

Gantt charts

A Gantt chart shows the utilization of resources over time, building up a detailed history of a simulation run. An example is shown in Fig. 5.5.

The utilization of, for instance, machines, labour and power can all be traced. Gantt charts are also used to show the sequence of jobs performed, for example the products that have been processed by a machine.

Unlike a time series, recordings are not made at regular time intervals but whenever some change takes place; the history is more detailed. However, since large quantities of information are produced, understanding and analysis are more difficult.

Pie charts

Pie charts show the proportional split of some value. Figure 5.6 is an example.

Effectively, they are a visual representation of the tabular reports discussed above. Typical uses are to show the proportions of a resource's use or the percentage split of customer types or product variants. Their

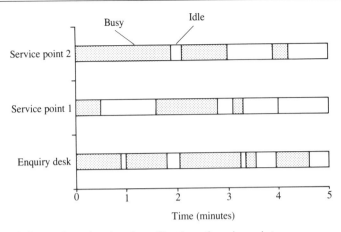

Fig. 5.5   A Gantt chart showing the utilization of service points

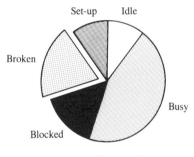

Machine utilization (OP10)

Fig. 5.6   A pie chart showing the percentage split of a machine's usage

shortcoming is that only an aggregate representation is given with no account of variability or time.

Scatter diagrams

A scatter diagram shows the relationship, or correlation, between two values. If the values are related then the points on the diagram form a relatively straight line, otherwise the plots are widely dispersed. Figure 5.7(a) shows an example of related data; the waiting time of a customer in a queue is related to the waiting time of the previous customer. Figure 5.7(b) shows an example of unrelated data; the length of a breakdown on a machine is not related to the previous repair time.

Scatter diagrams are particularly useful in testing assumptions about the relationship between two values. In this respect they act as an aid to model testing (Chapters 9 and 10). A common application is to ensure that the

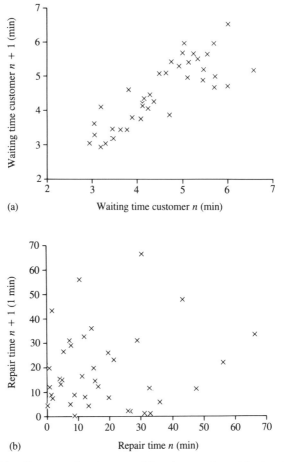

Fig. 5.7　Scatter diagrams of customer waiting times and machine breakdowns

occurrence of random events in a model does appear to be truly random (unrelated).

5.4　Viewing the reports

Having identified the values to be reported and their method of presentation, the final issue is how they should be viewed. There are four ways of viewing the reports and, in practice, it is likely that a mixture of these methods will be used:

1 *Dynamic display*　The tables and graphs can be displayed on the screen, changing as the simulation runs. These dynamic reports can prove very useful, especially for demonstrations and when the model is being used as

a training tool. The disadvantages are that there is often limited space on the screen and it is not always possible to print these reports, particularly to presentation standards.

2 *Interactive reports*   The provision of interactive menus, through which the reports are viewed, can overcome the difficulties of having a dynamic display. However, it is no longer possible to see the reports change as the model runs, which may limit its use for demonstrations and for training.

3 *Data files*   In both of the methods above, further manipulation of the results is not possible. Writing the results to external data files enables this to be done. It is also a useful means of maintaining a permanent record of a run.

4 *Third party software*   Having written the results to a data file, they could be imported to a spreadsheet, database or other third party software. This enables further analysis and the use of presentation graphics. It is important to ensure that the data file is in a format that can easily be read by the third party software.

Some examples of reports are shown in Table 5.2. The values to be reported and the methods of reporting and viewing are all given. Each is related to the examples of experimental factors given in Table 5.1. This is not intended to be an exhaustive list but a demonstration of the range of reports that might be given.

Conclusion

In this chapter the experimental factors (inputs) and reports (outputs or responses) have been discussed. Having identified these, it is time to consider the scope and level of the model, which is also known as conceptual modelling.

Summary

Experimental factors

- Identify the factors
- Determine the likely range of values
- Decide upon the method of data entry: the simulation model code, menu-driven options, data files, third party software

Reports: their roles

- Measuring the extent to which the objectives have been achieved
- Pointing to any problems and so enabling solutions to be identified

Table 5.2   Examples of reports

1 *Facilities planning*
  Objective: to improve customer service levels to 95% with the planned facilities

Values	*Reporting*	*Viewing*
Waiting time	Histogram, mean, standard deviation, minimum, maximum	Dynamic
Percentage service level	Cumulative total	Dynamic
Staff utilizations	Percentage, pie chart	Interactive

2 *Obtaining the best use of current facilities*
  Objective: to increase throughput by 10% within a capital spend of $100 000

Values	*Reporting*	*Viewing*
Daily throughput	Histogram, mean, standard deviation, time series	Dynamic, data file
Equipment usage	Percentage	Interactive, data file

3 *Developing methods of control*
  Objective: to select one of four methods of organizing labour that gives the highest throughput

Values	*Reporting*	*Viewing*
Hourly throughput	Time series, mean, maximum, minimum	Software
Labour utilization	Percentage	Software
Total jobs	Cumulative total	Software

4 *Materials handling*
  Objective: to determine the number of vehicles required to enable 100% availability of materials

Values	*Reporting*	*Viewing*
Material shortage	Per cent	Data file
Hourly shortage	Time series	Data file
Vehicle usage	Per cent	Data file
Hourly vehicle use	Time series	Data file

5 *Examining the logistics of change*
  Objective: to demonstrate that plant changes can be made with negligible interruption to manufacture

Values	*Reporting*	*Viewing*
Daily throughput	Histogram, mean, standard deviation, time series, maximum, minimum	Dynamic

6 *Company modelling*
  Objective: to reduce intersite movements by 20%, while maintaining profit levels

Values	*Reporting*	*Viewing*
Weekly movements	Time series, mean	Dynamic
Total movements	Cumulative total	Dynamic
Weekly profit	Time series, mean	Dynamic
Total profit	Cumulative total	Dynamic

7 *Operational planning*
  Objective: to identify the number of contract staff required in the coming month in order to achieve customer service levels

Values	*Reporting*	*Viewing*
Waiting times	Histogram, mean, standard deviation	Interactive
Percentage service level	Cumulative total	Interactive
Staff utilization	Percentage	Interactive

8 *Training operations staff*
  Objective: to train the supervisors in methods of selecting jobs so throughput targets can be attained

Values	*Reporting*	*Viewing*
Hourly throughput	Time series, mean	Dynamic
Total throughput	Cumulative total	Dynamic
Equipment usage	Percentage, pie chart	Dynamic

Reports: identification

- identify the values to be reported: based upon the objectives
- determine the method of reporting:
  - –Tabular reports: cumulative total and percentage, mean and standard deviation, median and quartiles, mode, minimum and maximum, statistical tests
  - –Graphical reports: time series, histograms, Gantt charts, pie charts, scatter diagrams
- Decide how the reports should be viewed: dynamic display, interactive reports, data files, third party software

Case study 5.1: Panorama Televisions

Experimental factors
Based upon the objectives of the project only two experimental factors are required:

- The number of pallets, with an expected range of 30 to 200, to be changed through the model code
- The size of buffers (conveyors), with a maximum 200% increase, to be changed through the model code

Reports
In order to report on the success of experiments and to identify possible problems, the following reports should be given:

- A time series of daily throughput, both total and for each product
- A histogram of total daily throughput
- The mean, standard deviation, maximum and minimum daily throughput
- Percentage machine utilizations, to include percentage idle, busy, blocked, broken and set-up

The first three reports will be part of the dynamic model display while the final report will be available through interactive menus. All the reports should be available for output to a data file so further analysis can be performed, possibly using third party software.

Notes
The aim of the first three reports is to enable a thorough analysis of daily throughput, both its distribution and behaviour over time. The final report is included to aid the identification of problems, for example an operation with significant stoppages.

Case study 5.2: Natland Bank

Experimental factors
Based upon the objectives of the project two experimental factors are required:

- The number of auto tellers, either one, two, or three, to be changed through menu-driven options
- Single queues or multiple queues, to be changed through menu-driven options

Reports
In order to report on the success of experiments and to identify possible problems, the following reports should be given:

- A histogram of waiting time for each customer, also including the mean, standard deviation, maximum and minimum waiting times
- The percentage of customers served within the three minute service level
- Percentage facility utilization, both in a table and a pie chart
- Cumulative totals showing the mean, maximum and minimum queue sizes and the total customer demand
- A time series showing the queue size and number of customer arrivals every 15 minutes

All reports should be available as part of the dynamic display to enhance the model's use in demonstrations.

Notes
Since the model is to be used as a communication tool and to convince managers of requirements the reports need to have a high visual impact.

The first three reports are central to the analysis of the model and aim to show the service level achieved and the resources required. Reports 1 and 2 show the service level in terms of the distribution of waiting times and the percentage of customers served within three minutes. The third report shows the utilization of resources required to achieve this service level. The final two reports give some additional information regarding customer demand and queues; their aim is to improve both understanding and communication.

6

The conceptual model: scope and level

When my client had completed her explanation of the manufacturing facility the white board was covered with drawings of machines, storage areas and methods of transportation. It had taken some time to grasp the detail of how it all worked, but in the end it was clear. Now we entered a discussion on how to approach a simulation.

'So, what are you going to include in the model?' I asked.

She looked at the board and thought for a moment. 'All of it,' she replied.

'All of it! Are you really sure you need to model all of it?'

Slightly perplexed she exclaimed, 'Well, I thought it was important to include everything in the model.'

'But didn't you say the problem was in the grinding area?' I asked, pointing to the bottom right corner of the white board.

'It is,' she replied.

'Surely then, you only need to model that.'

By now you could see the disappointment written across her face. After all, a simulation of the grinding area did not seem much of a challenge compared to a model of the whole facility. The fact was, however, that the time it would have taken to model everything did not warrant the minimal benefits it would have given.

In simulation projects there is a tendency to want to model 'everything' without stopping to consider exactly what is necessary. In the previous chapter the requirements for experimental factors and reports were discussed; in this chapter the needs of the model itself are covered. What should be included in the model? What can be left out? This is known as conceptual modelling or deciding upon the scope and level of the model.

Firstly, the terms scope and level are defined and then there is a discussion on how to form a conceptual model. Some common methods of scope and level reduction are discussed before the chapter concludes with the Panorama Televisions and Natland Bank case studies.

6.1   Scope and level of the model

SCOPE (WHAT TO MODEL?)
The scope is the range or the breadth of the model. In other words the question to be asked is:

• What should be included in the model?

At one extreme this could be the whole company, at the other, an individual operation or service point. In reality, the scope is often somewhere between these two.

LEVEL (HOW TO MODEL?)
The level is the amount of detail to be modelled or the model's depth. For each element within the scope the question should be asked:

• How much detail should be modelled?

The model may require a lot of detail or only the simplest representation. For example, when modelling an individual machine, should just the cycle time be represented, or should perhaps the operators, breakdowns, set-ups, repair labour, shift patterns and production schedules be modelled? Alternatively, some middle ground could be found. If a service point is being modelled, which out of random service times, staff rosters, queue jockeying and shortages should be included?

6.2   Conceptual modelling: deciding upon the scope and level

When considering what to include in a model the basic rule is:

• Model the minimum amount of detail required to achieve the project's objectives

The reasons for this are illustrated by the graphs in Fig. 6.1.

Figure 6.1(a) represents the expected accuracy of a simulation model in relation to its scope and level. Initially, increasing the scope and level leads to significant gains in accuracy. However, the advantage of further increases

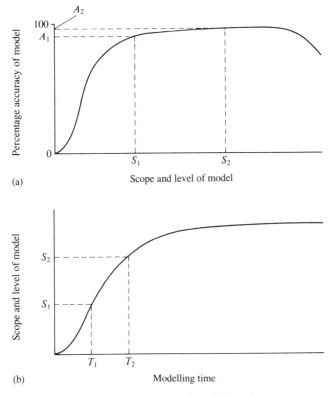

Fig. 6.1   Model accuracy, scope and level, and modelling time

is not as great; there are diminishing marginal returns. Basically, 80 per cent of the accuracy is obtained from 20 per cent of the model detail. Once too much detail is added, it is unlikely that enough good data is available to support the simulation and the accuracy of the model may actually be reduced—hence the dip in the curve.

Figure 6.1(b) shows the time required to build a model. Increments in the scope and level make the modelling process more complex and so an increasing amount of time is required to add more detail; there are diminishing marginal returns from spending more time over model building.

The graphs show that a small increase in the accuracy of the model from A_1 to A_2 requires an increase in the scope and level from S_1 to S_2. In turn, this means that the time required to build the model is approximately doubled from T_1 to T_2.

The principle is that if the model is too broad and too detailed precious time is wasted. On the other hand, if the model is too narrow and there is not enough detail then the accuracy is open to question. How can the right scope and level be determined?

The identification of the correct scope and level is very much an art, and an art that becomes easier with experience. Therefore, the employment of expert help is probably more useful at this stage in a project than at any other. However, since this is not always possible some guidelines are now given.

When forming the conceptual model, the scope and level, the following should be taken into account:

- The objectives of the project
- The time-scale, run speed, visual and interactive objectives
- The experimental factors and reports

These are now discussed.

THE OBJECTIVES OF THE PROJECT

The example given at the beginning of this chapter showed that it was not necessary to build a model of everything. The objectives were to overcome the bottlenecks in the grinding area and the scope could be quickly reduced to model only this. The key was to relate the scope and level of the model to the project's objectives.

When building a model, only the minimum scope and level should be included, but under the condition that the objectives of the project can be achieved. Some objectives require a very detailed model while others need only the simplest of simulations. In order to determine the scope and level the principle of successive inclusion can be used. Start with a basic conceptual model and gradually increase the scope and level. At each stage ask what impact the additional detail is likely to have on the model's ability to address the objectives. Once it is judged that increasing the scope and level has little impact on the objectives, stop. Alternatively, successive exclusion can be used, gradually removing details from a complex model until it is recognized that a significant impact is made.

Whichever method is adopted, a large amount of judgement is required. As a general rule, it is best to have a conceptual model with not enough detail than one with too much. The time required to build an overcomplex simulation cannot be reclaimed, but further detail can always be added to a simple model.

This is illustrated by a facilities planning project which I carried out with a manufacturing company. At the beginning of the project the objectives were discussed and it was clear that scheduling was perceived to be the problem. Since scheduling had an impact on the whole facility the scope of the model had to include everything. However, the level of detail could be reduced: the sequence of assembly had most impact on scheduling while the exact timings of each machine did not. Therefore, it was only necessary to model in detail sections of the facility where re-sequencing took place; other sections could be modelled using the 'black box' technique (Sec. 6.3).

On completion of the project, the objectives were achieved and it was shown that scheduling did not present a problem. However, in the process of simulating the scheduling strategy an indication of poor throughput was given. The model was not detailed enough to analyse this and a second project was raised to build a more detailed simulation. The first model focused on the scheduling objectives and made sure that an answer was given quickly. The second, more detailed, model could then focus on a new set of objectives related to throughput. A great deal of time could have been wasted if a single, very complex, model had tried to achieve both objectives from the outset.

THE TIME-SCALE, RUN SPEED, VISUAL AND INTERACTIVE OBJECTIVES
The discussion on objectives in Chapter 4 stressed the importance of setting time-scale, run speed, visual and interactive objectives. The scope and level of the model must also relate to these.

Some thought needs to be given to the scope and level that can be achieved within the time-scale. If a suitable model cannot be built, the time-scale or the objectives need to be changed.

The run speed should be considered; a more detailed model runs more slowly than a less detailed one. The scope and level should enable the model to run at the required speed.

A complex visual display normally requires a complex model to drive it. If the objective is to have a communication tool, it is quite justified to include details that are not strictly required other than for the purpose of the display.

Finally, the scope and level should allow for the interactive nature of the model. For example, if the objective is to change data part-way through a run, the model must be capable of accepting these changes.

THE EXPERIMENTAL FACTORS AND REPORTS
The experimental factors must be included in the model. There is no point in stating that breakdowns are an experimental factor if they are not part of the conceptual model. Neither should, for example, the manual tellers in a bank be excluded from the model's scope if they are to be used in the experiments.

The conceptual model must also be capable of providing the correct reports. If waiting times are a vital result, then the queues must be modelled in detail. If down times need to be reported, then breakdowns should be included.

COMPLETING AND VALIDATING THE CONCEPTUAL MODEL
Having completed the conceptual modelling process, it should be possible to form a list of elements to be included and the detail required for each. This list is then used as the basis for data collection, which is discussed in the next chapter. Any assumptions should also be noted as well as the basis for making them. Then the conceptual model needs to be validated in order to ensure that it is correct. A useful means for doing this is the project specification which is discussed in Chapter 8.

Throughout the process of conceptual modelling it is important to review the objectives of the project and ensure that they are realistic. Questions should be asked, such as:

- Can the objectives be achieved within the time-scale?
- Are the visual objectives realistic?
- Is it easy to represent an experimental factor?
- Does a particular report require an inordinate amount of modelling effort?

It is important to be flexible, to discuss alternative approaches and possibly to change some of the objectives. By iterating through the problem definition phase, challenging both the modelling approach and the objectives, the project has a greater chance of success.

6.3   Methods of simplification: scope and level reduction

Having given some guidance on forming the conceptual model, eight methods of scope and level reduction are now discussed; each is based on my own experience. The intention is to show some useful methods of simplification and not to provide an exhaustive list. It is hoped that these will act as a catalyst for further ideas.

Reducing the scope and level of a model often leads to some loss in accuracy. For each method discussed here it is useful to estimate the extent of this loss. A suggested method is to build two models showing a small section of the facility: the first, a detailed model, the second incorporating the reduced scope and level. By running the two models and comparing the results some indication of accuracy is given.

ONE ELEMENT TO REPRESENT A GROUP OF RESOURCES: 'BLACK-BOX' MODELLING
In 'black-box' modelling a section of a facility is represented as a time delay and no further detail is included. Black boxes are used to model anything

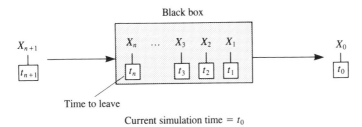

Fig. 6.2   Black-box modelling

from a group of machines or an enquiries area to a complete facility or plant. Elements that represent parts, customers, jobs or such like enter a black box and at some time later leave. This is shown in Fig 6.2.

As elements X enter the black box the time at which they are due to leave, t, is calculated. Once the simulation reaches time t the elements leave the box. The number of elements in the box at any one time is shown on the simulation display. The process can be extended to model re-sequencing (such as rectifications), stoppages and shift patterns by manipulating the time at which the elements are due to leave, the values of t.

The advantages are that it is a fast way to build a model and the simulation runs quickly. However, simplification inevitably leads to some loss in accuracy and the display is only basic. In a number of situations, however, these disadvantages are not important.

ONE ELEMENT TO REPRESENT A GROUP OF ITEMS

Instead of modelling individual items, an element can be made to represent a group of items. This is common practice in models of high-volume manufacture such as confectionery. For example, in a model of a wrapping line, ten chocolate bars are represented by one element and the machine cycles are multiplied by ten to suit.

The main benefit of this approach is that it reduces the number of events the model has to handle, which improves the run speed of the simulation. Also, it provides a way of using less computer memory, especially if a model is close to some hardware/software constraint. The main disadvantage is that characteristics, such as colour and size, can no longer be attributed to individual items.

EXCLUDING RESOURCES

The representation of labour, the most commonly modelled resource, can be simplified in two ways:

• Assume that labour is always available and so exclude it from the model

- Represent the labour element as an addition to the time required to perform an operation

For example, there is a significant difference between the repair time and down time for an operation. Repair time is purely the time required for repair once the labour has started work. Down time is the total time from failure to repair and includes the time spent waiting for labour. Therefore, if down time data are available the second method of simplification has already been achieved.

These methods are useful when there is a lack of good data, making it impossible to model labour exactly. These techniques can also be used when the involvement of labour has a negligible effect on performance. The shortcoming is that the detail of the interaction between the labour and the facility is lost.

Note that these approaches can also be used to model other resources such as power, fixtures and money.

EXCLUDING SHIFT PATTERNS

Including shift patterns in a simulation does not always add to the validity of the model. In general shift patterns should only be modelled when:

- Different areas work to different shifts
- The quantity of labour varies between shifts
- Some operations continue outside of the shift, for example, repairing broken machines
- Shifts are required to give credibility to the model

EXCLUDING SUB-COMPONENTS IN ASSEMBLY OPERATIONS

Most simulation models include some form of assembly operation. For example, in manufacturing, car engines are assembled to car bodies and sweet wrappers are assembled to sweets; in service operations, money is assembled to customers in a bank and food to a couple eating a restaurant meal.

In some situations many components are assembled and modelling them all would be very complex. By assuming that these components are always available they do not need to be modelled. This gives a simpler and faster simulation; however, when the results are analysed allowance needs to be made for this simplification.

As an alternative, a shortage of components can be modelled as a stoppage. By adding an allowance to the failure rates a realistic representation of stock-outs can be achieved.

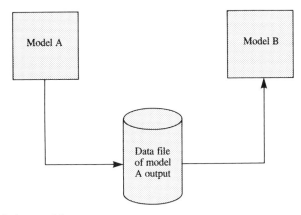

Fig. 6.3   Splitting models

SPLITTING MODELS

Sometimes a model can be split into two or more sub-models, the output of one sub-model being the input to another (Fig. 6.3). The models are linked by writing the output to a data file and reading this into the second model. The data file contains information on each element and the simulation time at which they were output.

The advantage of splitting models is that sub-models run faster and it is quite likely that the time to run every sub-model in series is less than the time required for an equivalent run with a single model. Also, if hardware/software constraints make it impossible to detail a facility in one model, then splitting the models is a potential solution. A final benefit is that models can be built in parallel, reducing the model building time.

The shortcoming of this approach is that the interference between the two models cannot be predicted easily. In other words, if there is a blockage in the second model its effect on the first model cannot be represented. For this reason, always aim to split a model at a point where there is a large amount of buffering, which minimizes the need to predict the interference effects.

EXCLUDING INFREQUENT EVENTS

In many situations there are a number of events that only take place at infrequent intervals. For example, a high bay crane in a storage facility may only break down once every two years. When such an event occurs some alternative operating procedure is normally implemented.

In general it is best not to include these events in a simulation model since they are not part of normal operating practice. If they are included, very long simulation runs are probably required in order to obtain a

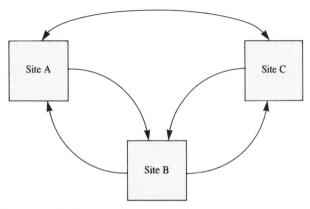

Fig. 6.4 Transportation between sites

reasonable result. As a rule of thumb, the length of the simulation run is sufficient when the most infrequent event has taken place on at least 10 to 20 occasions (see Sec. 11.4 on performing long runs).

An alternative approach to modelling infrequent events is to perform an experiment in which a specific event is made to occur. The effects of the event can then be discovered and various emergency procedures can be tested.

MODELLING DELIVERIES INSTEAD OF TRANSPORTATION
The movement of goods between sites is a common requirement in simulation (see Fig. 6.4). Modelling transportation such as forklift trucks, heavy goods vehicles or even trains can be complicated. Allowance may be required for factors such as breakdowns, punctures, traffic congestion, weather conditions, turnaround times and driver shifts.

As an alternative, the deliveries rather than the transportation can be modelled. Ask the questions: how many deliveries are made each day and what is the average movement time? The number of deliveries can then be timetabled to leave each site at either regular or random intervals and to arrive at their destination based on the movement times.

Although this approach is less accurate, it is much simpler. Another shortcoming is that the visual effect is not as great as a full model of the transportation.

Conclusion
By now the scope and level of the model should have been decided; a conceptual model has been built. The outcome is a list of elements to be modelled which in turn provides a list of the data required. These data must now be collected and analysed for use in the simulation.

Summary
Scope and level of the model

- Scope: the breadth, what should be included in the model?
- Level: the depth, how much detail should be modelled?

Conceptual modelling: deciding upon the scope and level

- Model the minimum amount of detail required to achieve the project's objectives
- Take into account: the objectives of the project; the time-scale, run speed, visual and interactive objectives; the experimental factors and reports
- Revisit the objectives if there are problems in building a model that achieves them
- Make a list of the elements to be modelled and any assumptions

Methods of simplification: scope and level reduction

- One element to represent a group of resources: 'black-box' modelling
- One element to represent a group of items
- Excluding resources
- Excluding shift patterns
- Excluding sub-components in assembly operations
- Splitting models
- Excluding infrequent events
- Modelling deliveries instead of transportation

Case Study 6.1: Panorama Televisions

Scope
The model need only include the television assembly plant specifically from OP10, load television to pallet, to OP60, unload television from pallet; this is shown in Fig. 6.5. The final test and pack need not be modelled since it is already known that there is enough capacity in this area.

Level
The following detail needs to be included in the model:

- General detail: production schedule, number of pallets
- Automatic operations: cycle times, breakdowns, set-ups on changeover, repair/set-up labour

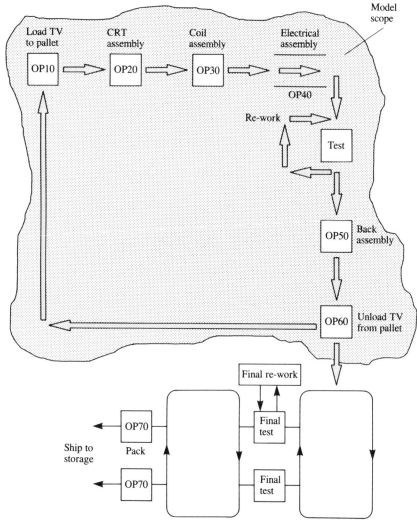

Fig. 6.5 Panorama Televisions: model scope

- Manual operations: cycle times
- Conveyors: capacity, transfer times
- Rework: percentage test failures, rework times

Assumptions
Stoppages (breakdowns and set-ups) need to be modelled in detail in order to represent the facility's main problem, namely interruptions to

manufacture through stoppages. The number of pallets and buffering between operations are also modelled since these are experimental factors.

The following are assumed:

- Conveyor breakdowns are infrequent
- Sub-components, such as television boxes and cathode ray tubes, are 100% available
- No work, including repairs to broken machines, takes place over a weekend

Therefore, none of these are modelled.

Case study 6.2: Natland Bank

Scope

The model need only include the operation of the auto tellers (see Fig. 6.6). An investigation of the manual tellers and the enquiry desk will be performed with two separate models which will not be built until the simulation of the auto tellers is complete, and only if there is time.

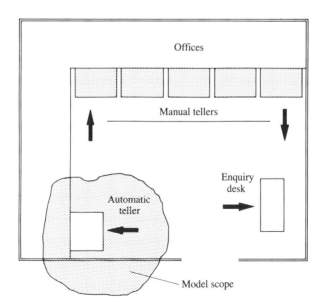

Fig. 6.6 Natland Bank: model scope

Level

The following detail needs to be included in the model:

- Customers: arrival rates
- Queues: single/multiple queues, queue discipline and jockeying
- Auto tellers: service times

Assumptions

It is assumed that auto teller failures are negligible and so they are not modelled.

7

Data collection and analysis

Once I drove all the way to South Wales for the sake of some data. The company had promised that all the information I required was available, and in some ways it was. I was greeted by an engineer who spent a number of hours showing me around the plant and presenting me with reams of information. It was all very interesting (so much that I did not even get a coffee!), but most of it was of little use for the purposes of the project. In fact, despite the glut of information, some of the data I required were not available, and much of the rest were in the wrong format to be used in the simulation. In the end I left, somewhat frustrated, and found the nearest 'watering hole' where I finally collapsed with a cup of coffee.

This is not an uncommon problem. Data are often not available and, even if they are, the information may either be inaccurate or in the wrong format. The purpose of this chapter is to show how the data requirements can be established and to discuss ways of dealing with unobtainable and inaccurate information as well as data that are in the wrong format. The problem of handling constant changes to the data is also discussed. Finally, the aim is to consider how to model randomness and select the correct distributions. Each of these subjects is discussed in turn and the chapter concludes by applying the concepts to the Panorama Televisions and Natland Bank case studies.

7.1 Data requirements
In the process of performing a simulation project various types of data are required. These are of a quantitative nature, for example cycle times, arrival rates and resource requirements; they explain logic rules such as the control of flows, scheduling strategies and work allocation; and they describe the physical layout. Some of these data are deterministic—their value docs not vary during a simulation run—while other data are stochastic—they are

subject to random variation during a run. Stochastic data are normally described using a distribution.

At this stage in the project it is useful to construct a list of all the data that need to be collected. Further to this, those responsible for gathering the data should be identified as well as the date by which the information needs to be available. This should obviously fall in line with the project time-scales.

The requirements can be determined by considering three specific uses of the data:

1 *To build the simulation model* The scope and level of the model, discussed in Chapter 6, provides a list of the elements and detail necessary to build the model. This is effectively a list of data requirements. It is important to be specific about exactly which data are required and to ensure that they are provided in the right format. For instance, it is best to ask for the average time between failures and average repair times, rather than just breakdown data.

2 *To set the initial level of the experimental factors* An estimate of the initial level of the experimental factors (Sec. 5.1) needs to be made. If the facility being modelled is already in operation then using data on current operating levels is most suitable. If the facility is new, modelling planned levels is best. This will aid the validation process (Chapter 10).

3 *To check the validity of the model* Once the model is built, its validity must be checked before any experiments are carried out. Therefore, it is important to have some data with which to perform this validation. For example, the results of a simulation could be compared with historic data on weekly throughput to ensure that it behaves in a similar manner. At this stage it is useful to identify the data needed for validation so they can be collected in readiness. The requirements for validation are discussed in detail in Chapter 10.

By considering these uses for the data it should be fairly straightforward to identify exactly what is required. These requirements can then be included in the project specification (Chapter 8) along with those responsible for collection and the date when the information needs to be available.

7.2 Methods of data entry
Another consideration is the method of data entry. This should suit the particular needs of the project and the model user. For example, data files are useful when large amounts of repetitive information are to be handled, enabling data to be edited on the screen and saved to separate files. The interactive objectives, discussed in the last part of Sec. 4.4, should act as an

aid when making this decision. As with the experimental factors in Sec. 5.1, there are four methods of data entry available:

- The simulation model code
- Menu-driven options
- Data files
- Third party software

The model user often needs to access data other than just the experimental factors. As a project progresses it is not unusual for the data to change and for new experiments to be identified. The user may also wish to access data in order to improve understanding of the simulation. The method of data entry and the flexibility of the model should allow for this.

7.3   Data categories

It is important to consider the availability of the data that are required. A useful framework is to categorize the data into one of three groupings shown in Table 7.1.

Category A represents those data that are immediately available. Common examples are machine cycle times, the physical layout and projected demand. However, these data are not always straightforward since they may be inaccurate or in the wrong format for the simulation. Methods of overcoming such problems need to be found.

Category B data are not available but can be collected within the available time-scale of the project. Control rules, production schedules and movement times often fall into this category. Collection can be performed by interviewing those with an expert knowledge of the facility, for example operators and supervisors. Alternatively, someone may need to physically stand by a process and watch what happens. Whatever method is used, it is important to ensure that the data are both accurate and in the correct format for the model, thereby ensuring that the simulation results are correct. Sometimes there is not enough time within the project to obtain a reasonable sample for the data. For instance, it can take three months to collect representative data on the breakdowns of a machine while a project may need to be complete in only one month. If the time-scale cannot be changed these data become category C.

Category C data are neither available nor can they be collected, normally

Table 7.1   Data categories

Category A	Available
Category B	Not available but collectable
Category C	Not available and not collectable

because of time and resource constraints, or simply because no similar process exists. Typical examples are machine breakdowns and customer service times. Unfortunately, category C data occur more often than is desirable and they are in fact present in most simulation projects. However, this does not necessarily jeopardize the project since there are a number of ways to deal with them.

The methods of dealing with unobtainable data are now discussed, followed by some ways of handling inaccurate information and data that are in the wrong format.

DEALING WITH UNOBTAINABLE DATA (CATEGORY C)

Unobtainable category C data may at first sight present a stumbling block to the success of the project. However, this is not the case. The following discussion shows four methods of dealing with data in category C.

Estimate the data

There are various ways of estimating data, for example:

* Studying similar facilities
* Interviewing operational staff
* Discussing the data with equipment vendors
* Making an intelligent guess

The particular method depends upon the circumstances of the project.

However, estimating the data presents a problem. No one is going to take seriously a recommendation that is based on uncertain data, nor is it a good idea to stake a career on the accuracy of a simulation result when the model is based on such data. How can this problem be overcome?

Firstly, when data have been estimated, always make clear the assumptions that have been made. Secondly, at the validation phase of the project (Chapter 10) perform some sensitivity analysis on these data. By changing their values the impact of any errors in the data can be understood; it may be that the simulation results are not sensitive. If the results are sensitive, it is important to provide the potential range of results by performing further sensitivity analysis during experimentation (Sec. 12.3).

Finally, aim to improve the data as the project progresses. It may not be possible to obtain absolutely accurate data, but some data collection may be possible. The value of collecting data beyond the end of the project should also be considered. The model could then be updated with this more accurate information and the validity of the project's results checked.

The model creates the data

Rather than asking what the data are, turn the question around and ask what the data need to be. In other words, use the model to create the data and then aim to achieve this in reality.

This approach was used when I was simulating a manufacturing facility for which there were no clear data on breakdowns. Although estimates had been made from similar facilities, there was a lot of uncertainty concerning their accuracy. It was impossible to improve the information since the facility was not yet in operation. Therefore, initial experiments were performed with the estimated data and then the question was asked: what is the maximum level of breakdowns that could be tolerated while still achieving target throughput? To answer this, the sensitivity of throughput to breakdowns was analysed. We then talked with the equipment vendors about the need for machine reliability. We also discussed which maintenance procedures were necessary in order to achieve the required reliability.

Simplify the scope and level of the model

The scope and level of the model may need to be simplified further in order to omit data that are not readily available. For example, if data cannot be obtained on the variation in cycle times then fixed times could be used. This approach is valid if omitting specific details does not significantly affect the accuracy of the model.

Change the project's objectives

As a last resort, the objectives of the project could be changed. For example, if one of the objectives is to consider transportation between two locations, but no data are available on transportation times, then this objective could be removed from the project. At a future date, when more information is available, the objective could be addressed properly.

The validity of this approach depends upon the importance of the objective. If a critical objective cannot be achieved then the project should probably be abandoned before any more time is wasted. In reality, this is unlikely to happen.

DEALING WITH INACCURATE DATA AND DATA IN THE WRONG FORMAT (CATEGORIES A AND B)

Data in category A initially appear to be straightforward enough. Even collecting category B data does not seem to present any real problem except for the time required to obtain the information. However, both the data that are to hand and the data that are collected must be accurate and in the right format for the simulation.

Data Accuracy

It is often said that a model is only as good as the data that are input to it, and this is certainly true. However, this does not automatically lead to the conclusion that all data must therefore be absolutely accurate; there are degrees of accuracy. Data that are used frequently in a model probably need to be more accurate than data that are not. For example, in a model of a bank, there are hundreds of customers arriving every hour while the automatic tellers probably fail on only a few occasions in a week. A 10 per cent error in the customer arrival rate is likely to have a greater impact on the model's accuracy than a similar error in the data on teller failures.

The principle is that there is little point in spending large amounts of time improving the accuracy of data that are used infrequently while there are errors in the more frequently used information. Frequent errors are likely to have a greater impact on the overall model accuracy. Time should first be invested by improving the frequently used data and secondly by considering information which is used less.

If the available or collected data are not considered to be sufficiently accurate then what can be done? The first option is to accept the data but to recognize the inaccuracy when analysing the results and making recommendations. It is necessary to judge the overall accuracy of the simulation model. This does not give a great deal of confidence and should probably be viewed as a last resort.

A second approach is to use both judgement and analysis to establish the true values of the data. For instance, an expert in the operation of the facility may be able to give some insight into the direction and magnitude of any changes that should be made.

If neither of these methods is appropriate then it is probably best either to collect the data again, possibly from an alternative source, or to consider it to be unobtainable category C data. Since it may be impossible to find an accurate source for the information, the latter may be the only option.

Data format

A common problem is that the data are not in the correct format for the simulation model. A typical example is machine breakdowns: some simulation packages model the time between failures as the operating time between failures (excluding idle and off-shift time) while data are often collected on absolute time between failures. Firstly, it is important to know how the simulation software uses the data and, secondly, it is important to ensure that the data are in the correct format.

Data that are in the wrong format can be handled in a similar fashion to inaccurate data:

- Accept the wrong format and make judgements about the simulation results
- Use judgement and analysis to convert the data to the right format
- Collect the data again
- Treat the data as unobtainable category C

7.4   Handling data changes

Many simulation projects are carried out in an environment of change, especially when a new facility is being modelled. Adjustments are constantly being made to the layout, timings, control and other data. A lot of time is spent keeping up to date with the changes and including them in the model; it is not long before the time-scales of the project begin to slip.

In order to avoid this situation, it is important to have a procedure in place for handling changes to the model data. A useful means for doing this is to agree a time after which the data will be frozen and no further changes to the model will be made; a date near to the completion of model building is probably best. By doing this, the time-scales of the project are more likely to be kept.

However, the real world continues to progress despite the frozen data in the model and so it must be recognized that there will be some inaccuracy in the simulation result. In general, the benefit of keeping to time is greater than the disadvantage of the inaccuracy, as long as it remains small. There are situations where the changes are so significant that it is necessary to unfreeze the data. In one project, with which I was involved, a new production facility was being modelled. Part-way through the project a complete re-design of the plant was made and it would have been meaningless to continue with the initial simulation.

A procedure needs to be put in place for recording any changes to the data after the cut-off date, thereby keeping a note of the variance between the model and reality. This information could be used to judge the expected inaccuracy of the simulation result. Alternatively, at a certain point in the project the model could be updated to reflect all these changes. This could be done, for example, just before the model is validated, for a second phase of experimentation, or as an additional project.

7.5   Modelling randomness

One of the main reasons for using simulation is its ability to model random events. Much of the data that are collected reflect this randomness, for example arrival rates, repair times, rectification rates and travel times. Having collected these data, it is important to ensure that the random nature of the facility is modelled correctly. In this section the methods of modelling random events are discussed, namely data streams and distributions. The different types of distribution data are then described. Finally, there is an overview of the main statistical distributions and a discussion on how to select the correct one.

METHODS OF MODELLING RANDOMNESS

Random events can be modelled in three different ways:

1 *Data streams* A data stream is a list of events that take place and is normally held in some form of data file. For example, a record could be kept of the times at which cars arrive at a petrol station (see Table 7.2). These times can then be replayed to the model in order to imitate exactly the arrival pattern of the vehicles. A flight departure and arrival schedule at an airport is another example of a data stream.

2 *User defined distributions* A user defined distribution shows the frequency with which various data values occur. This is normally based on some historic data that have been collected. For instance, the car arrival pattern has been summarized with a distribution and is shown in Fig. 7.1.

3 *Standard statistical distributions* There are a number of standard statistical distributions that can be used to describe the random nature of events. These are discussed in detail later in this section. An example is shown in Fig. 7.2. A negative exponential distribution has been used for the interarrival time of cars at the petrol station.

The method selected depends largely upon the situation and the advantages and disadvantages of each approach.

Table 7.2 Data stream: car arrivals at a petrol station

Car number	Arrival time (*hours:minutes.seconds*)
1	10:02.12
2	10:02.27
3	10:02.56
4	10:03.17
5	10:03.19
.	.
.	.
.	.

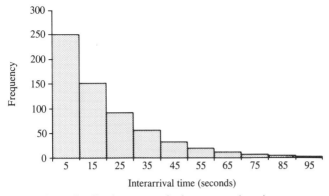

Fig. 7.1 User defined distribution: car arrivals at a petrol station

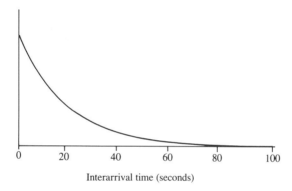

Fig. 7.2 Standard statistical distribution, negative exponential: car arrivals at a petrol station

Data streams give the significant benefit that a particular set of circumstances can be modelled exactly. For instance, if the purpose of the simulation is to model a specific period of operation then a data stream is a useful means for doing this. The shortcoming is that the data are limited to the sample that has been collected; in reality the range of the data may be greater than represented by the data stream. Another disadvantage is that only one set of circumstances can be modelled unless additional data streams are available.

A user defined distribution overcomes the second shortcoming. By running a number of simulation experiments, taking different samples from the distribution each time, a variety of circumstances can be modelled. However, the range of the data is still limited to the sample that has been collected.

Standard statistical distributions do not limit the range of the data. A distribution is fitted to the data (see the last part of this section) which represents all the possible outcomes and not just a sample. The main shortcoming is that if the base data are particularly skewed, it is probably impossible to fit a statistical distribution and a user defined distribution should be used instead. For example, a bimodal distribution, shown in Fig. 7.3, cannot be represented easily with a statistical distribution and so a user defined distribution should be used instead.

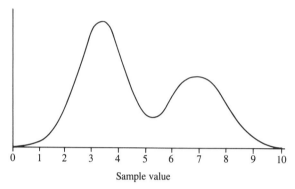

Fig. 7.3 A bimodal distribution

As a general rule, unless circumstances dictate otherwise, statistical distributions should be used in favour of user defined distributions which in turn should be used in preference to data streams.

TYPES OF DISTRIBUTION DATA
Distributions can be made to represent integer or real values and discrete or continuous data. Most simulation software provides these as options and it is important to select the correct one.

Integer or real

Integer values are whole numbers such as 0, 1, 2, 23, 1045 and −35. An integer distribution should be used when it is impossible for anything other than a whole number to be sampled. For instance, integer distributions are used to determine the number of operations between failure and the number of rejects in a batch.

Real values have decimal places such as 1.045, 20.4, 1723.3546, −45.34 and 1.00. If there is potential for a value to take a decimal place, a real distribution should be used. In general, if a distribution represents time, for example service times or cycle times, it is real.

Discrete or continuous

If a distribution is discrete then only the data supplied can be sampled from it. For example, a discrete distribution of service times may state the following:

Service time (min)	Probability (%)
1.5	20.0
2.8	50.0
4.0	30.0

In this case the service time can only ever be 1.5, 2.8 or 4.0 minutes.

In a continuous distribution, data are provided in a range and sampled accordingly. In the example above, a continuous distribution could be as follows:

Service time (min)	Probability (%)
0.0–1.5	20.0
1.5–2.8	50.0
2.8–4.0	30.0

Therefore, service times of 1.2, 3.4 and 2.7 minutes could be sampled from the distribution.

It is possible to have any combination of integer/discrete, integer/continuous, real/discrete and real/continuous distributions. The selection of the correct type depends upon the circumstances in which it is to be used. In general, but not always, distributions that sample times are real/continuous and distributions that sample quantities are integer/discrete.

STATISTICAL DISTRIBUTIONS

In this section, 11 statistical distributions are described. Normally they are provided as standard options in the simulation software. Of these distributions the most commonly used are the negative exponential, Erlang, gamma, normal, triangular and uniform. For each one the following details are given:

• Parameters
• Typical applications
• Shape
• Additional comments

Beta

Parameters: shape, scale
Applications: time to complete a task, modelling proportions (e.g. proportion of defects in a batch)
Shape (Fig. 7.4):

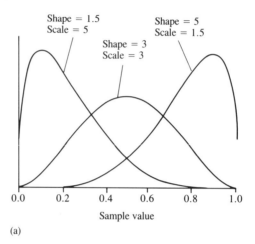

(a)

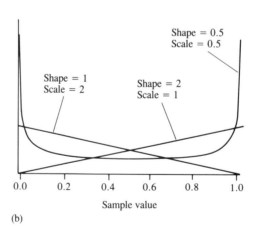

(b)

Fig. 7.4

Comments: The beta distribution is particularly useful as a rough model in the absence of data.

Binomial

Parameters: probability, trials
Applications: total successes in a number of trials (e.g. number of defects in
                 a batch)
Shape (Fig. 7.5):

Probability = 0.1, trials = 10   Probability = 0.5, trials = 10

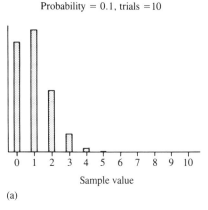 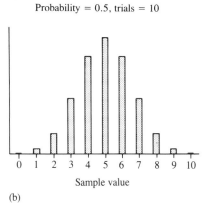

Sample value   Sample value

(a)   (b)

Fig. 7.5

Comments: The binomial distribution samples discrete data only.

Erlang

Parameters: mean, K value
Applications: time between events (e.g. interarrival time, time between
                 failure), time to complete a task (e.g. service time, repair
                 time)
Shape (Fig. 7.6):
Comments: The Erlang distribution is a special case of the gamma distribu-
                 tion. The K value is equivalent to the shape parameter in the
                 gamma distribution and the mean is equivalent to the scale.
                 For example, the Erlang mean X, K value Y is the same as
                 the gamma shape Y, scale X. However, the K value can only
                 be an integer while the shape parameter can be a fraction,
                 giving the gamma distribution a wider range of shapes.

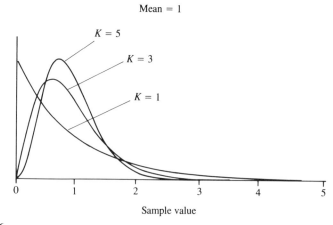

Mean = 1

K = 5

K = 3

K = 1

Sample value

Fig. 7.6

Gamma

Parameters: shape, scale

Applications: time between events (e.g. interarrival time, time between failure), time to complete a task (e.g. service time, repair time)

Shape (Fig. 7.7):

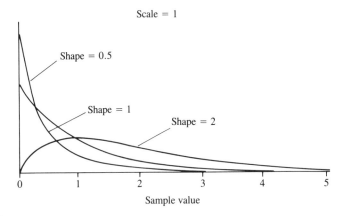

Scale = 1

Shape = 0.5

Shape = 1

Shape = 2

Sample value

Fig. 7.7

Comments: 1 This distribution has the widest range of shapes for modelling time between events and time to complete a task.

2 The Erlang and negative exponential distributions are special cases of the gamma distribution.

Log normal

Parameters: mean, standard deviation
Applications: time to complete a task
Shape (Fig. 7.8):

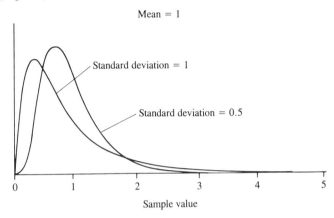

Fig. 7.8

Comments: This is a similar shape to the gamma, Weibull and Erlang
distributions but the 'hump' is higher.

Negative exponential

Parameter: mean
Applications: time between events (e.g. interarrival time, time between
failure), time to complete a task (e.g. service time, repair
time)
Shape (Fig. 7.9):

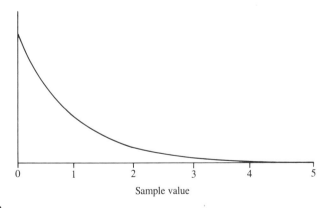

Fig. 7.9

Comments: 1 When only mean times are known, this distribution is a useful approximation.

2 The negative exponential distribution is a special case of the Erlang, gamma and Weibull distributions. The negative exponential mean = X is the same as:

Erlang mean = $X, K = 1$
Gamma shape = 1, scale = X
Weibull shape = 1, scale = X

Normal

Parameters: mean, standard deviation
Applications: errors (e.g. measurements of size and weight), dimensions (e.g. size and weight)
Shape (Fig. 7.10):

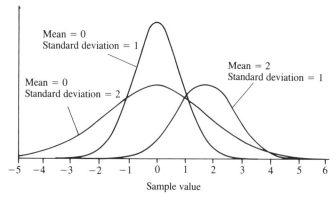

Fig. 7.10

Comments: The normal distribution is typically used for modelling quality inspection and control.

Poisson

Parameter: mean
Applications: number of events in a period of time (e.g. number of operations between failure), number of items in a batch (e.g. the number of boxes on a pallet of random size)
Shape (Fig. 7.11):
Comments: The Poisson distribution samples discrete data only.

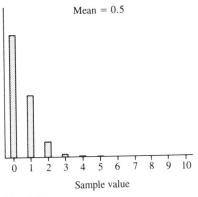

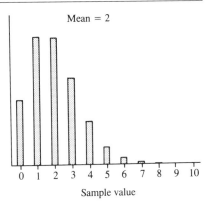

Fig. 7.11

Triangular

Parameters: minimum, mode, maximum
Applications: a first-pass distribution
Shape (Fig. 7.12):

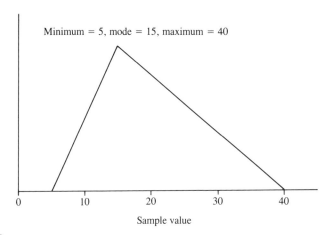

Fig. 7.12

Comments: A useful first-pass distribution in the absence of data. For
            example, when collecting data on repair times ask for the
            most likely time, the minimum and maximum and then use a
            triangular distribution.

Uniform

Parameters: minimum, maximum
Applications: generation of random variables for sampling, a first-pass
distribution
Shape (Fig. 7.13):

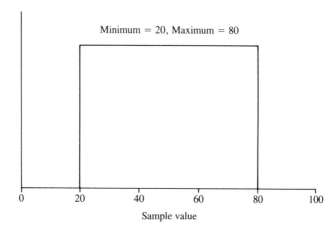

Minimum = 20, Maximum = 80

Sample value

Fig. 7.13

Comments: 1 Random variables can be generated for a customized
algorithm, for example, randomly selecting a part with
equal probability from one of three storage areas.
2 When little is known about a distribution this is a useful
first-pass representation of the data.
3 The uniform distribution can sample both discrete and
continuous values.

Weibull

Parameters: shape, scale
Applications: time between events (e.g. interarrival time, time between
failure), time to complete a task (e.g. service time, repair
time)
Shape (Fig. 7.14):
Comments: Reliability issues are often modelled with the Weibull distribution, for instance the time between failures.

PSEUDO-RANDOM NUMBER STREAMS
When building the model, the distributions normally require an additional
parameter, a pseudo-random number stream. A random number is a value

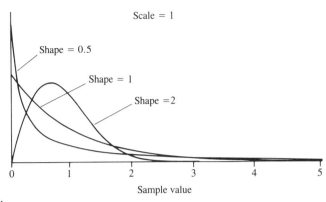

Fig. 7.14

that is obtained in a purely random fashion, that is with no particular pattern. Everyday examples are tossing a coin, giving a 'head' or 'tail', and throwing a die, giving a number between one and six. The main properties of random numbers are that:

1 There is an equal probability of any number being generated, i.e. when rolling a die a value of one is as likely as a value of two, etc.
2 The numbers are completely independent, i.e. the fact that a six was rolled last time does not affect the probability of a six being rolled again.

When using a computer, random numbers are generated by special mathematical techniques which normally provide values between 0 and 1 or between 0 and 100. The main property of these generators is that the same pattern of numbers is always generated; this is a particularly beneficial feature when performing simulation experiments (Sec. 9.2, model coding). These techniques are said to provide pseudo-random numbers since their values have all the properties of random numbers but they are generated in a mathematical and not truly random fashion. Most simulation packages provide more than one set of pseudo-random numbers and each set is referred to as a pseudo-random number stream. The use of pseudo-random number streams is discussed further in Sec. 9.2.

SELECTING THE CORRECT STATISTICAL DISTRIBUTION
Unfortunately, only on extremely rare occasions does some data present itself as, say, a Weibull distribution with the shape and scale parameters already defined. Indeed, one of the jobs of the data provider is to analyse the data and draw conclusions about their statistical distribution and parameters, a process which is discussed in this section.

Distributions need to be selected under two different circumstances:

- In the absence of any data regarding the distribution's shape
- When data are available on the distribution's shape

Before discussing these it is worth noting that there are a number of software packages available that can fit a distribution to some data. Two examples are UniFit II and SIMSTAT, neither of which are particularly expensive.

In the absence of any data regarding the shape of the distribution simple rules of thumb must apply. For example, a customer arrival rate is often expressed as a mean average with no reference to the shape of the distribution. In this case, it is probably best to use a negative exponential distribution since it is known to describe interarrival times. To improve the accuracy of the data, a number of different distributions could be shown to an expert in the facility being modelled. They would then be asked to select the one that most closely matched their view of the data. In the example above, an expert could be shown some Erlang distributions with various K parameter values.

However, there are occasions when information regarding the shape of the distribution is available. This is typically presented in the form of a histogram. The data are split into ranges and the frequency with which the data falls into each range is recorded. The aim then is to fit a statistical distribution to these data. The process of selecting the correct distribution and its parameters can be performed in three steps:

- Select a statistical distribution
- Estimate the parameters
- Check how representative it is

This is an iterative process in which the selection of a distribution and its parameters are gradually improved. The process is now explained by use of an example.

Some data have been collected on the repair time of a machine which is summarized in Table 7.3 and shown in Figure 7.15. A statistical distribution is now fitted to these data.

Select a statistical distribution

The first step is to select a statistical distribution. By visually inspecting the shape of the data it is possible to draw conclusions about a probable candidate. Initially, an Erlang distribution is used since it appears to match the shape of the data.

Table 7.3 Example: repair time data

Repair time (min)	Frequency
0.0–3.0	10
3.0–6.0	37
6.0–9.0	23
9.0–12.0	14
12.0–15.0	7
15.0–18.0	5
18.0–21.0	2
21.0–24.0	2
24.0–27.0	0
27.0–30.0	0

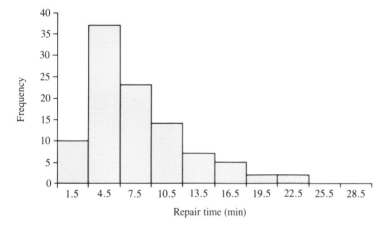

Fig. 7.15 Example: repair time distribution

Estimate the parameters

Secondly, the parameters of the distribution must be estimated. For parameters such as the mean and standard deviation it is normally possible to calculate them. Visual inspection and estimation are the best methods for selecting parameters that cannot be calculated, for instance the shape and scale.

The mean of the distribution is calculated in Table 7.4. Since it has been necessary to calculate the mean using the mid-point in each range, it is only an estimate. Whenever possible it is best to calculate a more exact value from the original data. For now, K parameters of one, three and five are used.

Table 7.4 Mean average of repair time data

Mean repair time (min)	Frequency	Mean repair time × frequency
1.5	10	15.0
4.5	37	166.5
7.5	23	172.5
10.5	14	147.0
13.5	7	94.5
16.5	5	82.5
19.5	2	39.0
22.5	2	45.0
25.5	0	0.0
28.5	0	0.0
Total	100	762.0

$$\text{Mean average} = \frac{762}{100} = 7.6 \text{ minutes}$$

Check how representative it is

The final step is to check how representative the selected distribution and its parameters are. For now, this is carried out by visually comparing the collected data with the expected frequencies from the proposed distributions. However, there are statistical tests for performing this analysis which are discussed in Sec. 14.2.

The expected frequencies need to be generated. In some cases these can simply be calculated, for example the samples from a uniform distribution; for other distributions this is not as straightforward. A practical method is to generate the expected frequencies in the simulation software. By taking a large number of samples, say 100 000, from the proposed distribution and recording them in a histogram with the same ranges as the collected data a good approximation can be made. The results for the three Erlang distributions are shown in Table 7.5. Note that the data sum to a hundred in order to make comparison with the data in Table 7.3 possible.

Histograms comparing the collected data with each of the expected frequencies are shown in Fig. 7.16. The Erlang distribution with a mean of 7.6 and K parameter of three most closely matches the information.

Having selected a distribution, the aim should be to improve the fit by iterating through this selection process. In the example, other K parameters could be tested and another distribution could be tried, for instance a Weibull.

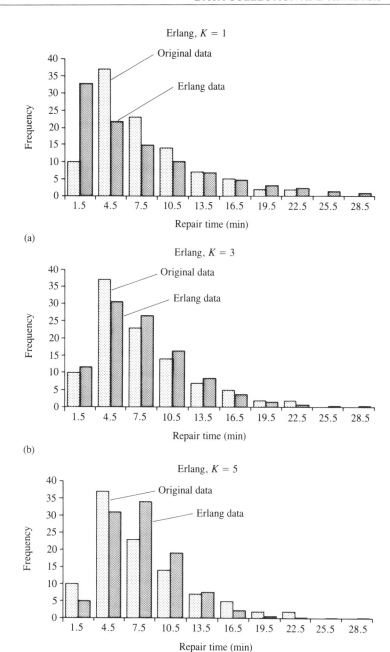

Fig. 7.16 Visual inspection of repair time distribution fit

Table 7.5   Expected frequencies for Erlang distributions

Repair time (min)	Expected frequency Erlang (7.6, 1)	Expected frequency Erlang (7.6, 3)	Expected frequency Erlang (7.6, 5)
0.0–3.0	32.8	11.6	5.0
3.0–6.0	21.7	30.6	31.1
6.0–9.0	14.8	26.6	34.2
9.0–12.0	10.1	16.4	19.0
12.0–15.0	6.7	8.5	7.6
15.0–18.0	4.5	3.7	2.3
18.0–21.0	3.1	1.6	0.6
21.0–24.0	2.0	0.7	0.2
24.0–27.0	1.3	0.2	0.0
27.0–30.0	0.9	0.1	0.0
> 30.0	2.1	0.0	0.0
Total	100.0	100.0	100.0

Selection of cell widths

In the example presented here the data were already represented by a histogram. If this is not the case a histogram needs to be constructed from the base data. In this process a decision has to be made concerning the width of the ranges in each cell; this was set at three minutes in the example. It is recommended that a number of different cell widths are tried until a smooth histogram is obtained. Figure 7.17 shows an example of a histogram constructed from the same base data as the repair time example but with a cell width of one minute. Whereas a smooth histogram was obtained in Fig. 7.15 with a cell width of three minutes, the distribution in Figure 7.17 is not smooth. Therefore, the original histogram is preferred.

Conclusion

In this chapter, the details of collecting and analysing the data have been discussed. Since this may take some time it is often performed in parallel with the other modelling activities. The definition of the problem is almost complete except for communicating its details and gaining agreement to proceed. A project specification is the means for doing this.

Summary

Data requirements

- Data types: quantitative, logic rules, physical layout, deterministic, stochastic

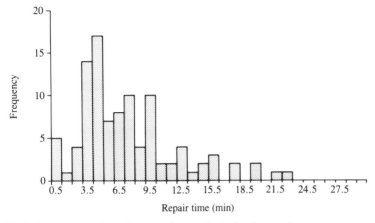

Fig. 7.17   Histogram of repair time with a cell width of one minute

- Required to: build the simulation model, set the initial level of the experimental factors, check the validity of the model

Methods of data entry

- The simulation model code
- Menu-driven options
- Data files
- Third party software

Data categories

- Category A: available
- Category B: not available but collectable
- Category C: not available and not collectable
- Dealing with unobtainable data: estimate the data, the model creates the data, simplify the scope and level of the model, change the project's objectives
- Dealing with inaccurate data and data in the wrong format: accept the data and make judgements about the simulation results, improve the data with judgement and analysis, collect the data again, treat the data as unobtainable category C

Handling data changes

- Agree a date to freeze the data
- Unfreeze the data for significant changes
- After the model data have been frozen, do keep a record of any data changes
- Update the model at specific points to reflect data changes

Modelling randomness

- Methods of modelling randomness: data streams, user-defined distributions, standard statistical distributions
- Types of distribution data: integer or real, discrete or continuous
- Statistical distributions: beta, binomial, Erlang, gamma, log normal, negative exponential, normal, Poisson, triangular, uniform, Weibull
- Selecting the correct distribution in the absence of data: rules of thumb, visual inspection
- Selecting the correct distribution when data are available: select a statistical distribution, estimate the parameters, check how representative it is

Case Study 7.1: Panorama Televisions

The actual data for the model are listed in Appendix 2.

Data requirements

DP = data provider

Data required	*Responsible*	*Available by*
Physical layout	DP	15 March
Production schedule:		
Batch quantities	DP	16 March
Production sequence	DP	16 March
Number of pallets	DP	15 March
Machines (OP20, OP30, OP50):		
Cycle times	DP	15 March
Breakdowns:		
Time between failure	DP	16 March
Repair time	DP	16 March
Number of repair labour	DP	16 March
Set-ups on changeover:		
Set-up time	DP	16 March

Manual operations (OP10, OP40, OP60):

Cycle time	DP	16 March
Number of stations at OP40	DP	16 March

Conveyors:

Capacity	DP	16 March
Transfer times	DP	16 March
Conveyor type	DP	16 March

Rework:

Percentage test failures	DP	16 March
Rework times	DP	16 March
Total number of repair/set-up labour	DP	16 March

Validation data:

Mean daily throughput	DP	16 March
Distribution of daily throughput	DP	16 March

After 16 March no further changes will be made to the data in the model.

Case Study 7.2: Natland Bank

The actual data for the model are listed in Appendix 2.

Data requirements

DP = data provider

Data required	*Responsible*	*Available by*
Physical layout	DP	14 November

Customers:

Arrival rate at auto tellers	DP	14 November
Queue discipline	DP	14 November

Auto tellers:

Service times	DP	14 November

After 14 November no further changes will be made to the data in the model.

Validation data
No validation data are requested since this is a model of a new facility and no data for comparison are therefore available.

8

The project specification

As children, and even as adults, most of us have played Chinese whispers. Everyone sits in a circle and someone is nominated to pass on a message which is then whispered around the circle until eventually it returns to its originator. Having passed through a number of people's lips and having been interpreted in various ways the message is almost bound to be completely different from its original form. For example, the message might start out as 'I would like a hamburger for tea tonight' but on return to the originator it may well be something like 'I got up an hour late this morning'. The interesting part of the game is attempting to discover at which points the message became confused and why.

Much the same effect takes place within business organizations. The simplest of messages can soon alter and be changed beyond recognition, rumours start and ideas get blown out of proportion. Equally, this can occur with a description of the objectives and scope of some simulation work. The original message might be that the simulation is going to help improve throughput; however, soon it is being said that there will be reductions in labour, scheduling policies created and major capital savings. Since unrealistic expectations are therefore raised, this leaves the simulation team with an impossible task when it comes to achieving success. The answer: write the message down and do not leave it solely to word of mouth.

However, a written message is also open to misinterpretation. In 1854 the British army faced the Russians at Balaclava. General Airey wrote a message for the Light Cavalry Brigade who proceeded to follow what they believed to be the orders. They charged the length of a valley under artillery and rifle fire, and, suffice it to say, few survived. The death of many soldiers, and the famous Charge of the Light Brigade, had been caused by a poorly written message which had been interpreted wrongly. The moral is to ensure that the message is not only written down but also that it is written down clearly.

The aim of this chapter is to emphasize the need to clearly communicate the objectives and scope of a simulation project and to show how this can be achieved with the provision of a project specification. The following are discussed: the reasons for providing a project specification, its structure, methods of communication and the problem of handling changes to the specification. The chapter concludes with example specifications for the Panorama Televisions and Natland Bank projects.

8.1 Why provide a project specification?

The project specification, or terms of reference, is the final stage in the problem definition phase. Its aim is to obtain a consensus and agreement for the way in which the problem has been defined and the project will be conducted. The objectives of the project have been set, the modelling approach determined and the data requirements identified. The specification summarizes each of these and should be communicated in order to obtain comment on its accuracy. Any requirements for change should then be included in the problem definition, the specification re-issued and further comment sought. This iterative process should continue until there is a consensus that the specification is correct.

There are five specific reasons for providing a specification.

TO COMMUNICATE THE PROJECT'S OBJECTIVES AND APPROACH
Every member of the project team needs to know, and understand, the objectives of the project and the method of approach. Without this, the success of the project may be threatened through a lack of clear direction and through misunderstandings. You may know what you are doing, but does anybody else? A project specification is a useful means for meeting these needs.

TO CONFIRM THE VALIDITY OF THE APPROACH
The project specification is necessary to confirm the validity of the approach and more specifically that:

- The objectives of the project are correct
- The right experimental factors and reports have been identified
- The modelling approach, the scope and level, is valid

By clearly communicating the specification of the project and obtaining feedback, it is possible to judge the validity of the intended approach. This is referred to as testing the validity of the conceptual model.

TO ENSURE CREDIBILITY
Validity ensures that the specification is correct; credibility ensures that the simulation is trusted. Although a model may be totally valid, if there is no confidence in the results it is of little use. Providing a specification is part of the process of winning people's confidence. By including them in the initial stages of modelling they are able to provide input and discuss any reservations. They feel that their opinions are of value and that these have been taken into account. This takes advantage of the fact that people are far more ready to trust their own ideas than those which are imposed on them by others.

TO REDUCE THE NEED FOR CHANGES TO THE SIMULATION MODEL
Having communicated the specification and received feedback, a number of problems may be identified. These can then be addressed before the simulation model is built. The benefit is that any omissions can be added to the specification and so potentially complex and lengthy changes, which would otherwise need to be made later in the project, can be avoided. This saves time and helps to ensure that the project meets its objectives. Although the expectation is that further changes will almost certainly be identified as the project progresses (Sec 8.4), the specification acts as an initial audit of ideas.

AS A REFERENCE DURING AND AFTER THE PROJECT
In order to meet the objectives of the project and keep to the time-scale it is important to ensure that the work remains on track. The specification acts as a focal point. Throughout the study it can be referenced in order to ensure that the work is progressing in the right direction and at the right speed. Once the project is complete, the success of the work can be measured against the detail of the specification.

8.2   The structure of the specification

The specification is a management summary of the project's objectives and the intended method of approach. It is not meant to be a full-blown technical specification including details of intricate modelling techniques. As a guide, the following contents are suggested; however, some sections may be omitted and others added depending on the needs of the project and the organization:

- Introduction to the problem and project objectives (Chapter 4)
- Expected benefits (Sec. 2.6)
- Model summary: scope and level of detail, assumptions, experimental factors and reports (Chapters 5 and 6)

- Data requirements: what is required, who is to provide the data, when it is to be available, when the data is to be frozen (Chapter 7)
- Time-scale and milestones (Sec. 3.2)
- Estimated cost (Sec. 2.6)

This information is discussed in the chapters or sections shown in brackets. At this stage the estimates of time-scale and milestones should be more thorough than those discussed in Sec. 3.2. These should include expected completion dates for major phases of the project, notably model building and testing, experimentation and reporting. For larger projects it may be useful to give additional milestones which help to keep a closer track of progress; for example model building could be split into a number of stages.

8.3   Communicating the specification

The specification needs to be communicated to each member of the project team who in turn should ensure that all other interested parties are informed. This includes any relevant managers and those who are represented by members of the project team. In this way all interested parties are given the opportunity to give feedback.

The project specification can either be written or verbal, or both. On smaller projects an informal verbal presentation may suffice, while on larger projects a written specification, possibly in combination with a formal presentation, is probably best. The method of communication also depends upon the culture of the organization. Whatever method is chosen, the ultimate aim is to obtain agreement for the specification.

WRITTEN SPECIFICATIONS

The advantage of a written specification is that it forms a permanent record of the problem definition. However, merely sending it out does not guarantee that it will be read; nor does it always command a response. Some effort may be required to ensure that it is read and that feedback is obtained, possibly by giving cut-off dates for comment and pro-actively chasing a response. Alternatively, presenting the specification as well as providing a written version may be advantageous.

The length of the specification should be kept to a minimum and in most cases a maximum of 10 pages should suffice. By keeping it short there is a greater chance that the specification will be read.

FORMAL PRESENTATIONS

When presenting the specification it is important that all parties are present in order to ensure that everyone has an opportunity to give feedback. It may be necessary to run more than one session, although this leads to some problems in co-ordination and so is probably best avoided whenever possible. At the end of the presentation, opportunity should be given for discussion, which may result in some changes to the specification. If possible, agreement should be reached by the end of the session in order for the work to progress.

The main advantage of a presentation is that it guarantees that all parties have at least heard the specification and provides an opportunity to obtain immediate feedback. Another advantage is that an open discussion helps in gaining a consensus of opinion, although this discussion needs to be handled well. However, if the presentation is made in isolation, without a written specification, then there is no formal record for reference later in the project.

VERBAL SPECIFICATIONS
On small projects a verbal specification may suffice. This is simply a case of discussing the needs of the project in an informal manner, even over a cup of coffee. It is recommended that a summary of the discussion is provided on a single sheet of paper to act as a more permanent record.

8.4   Handling changes to the specification
As a project progresses the requirements often change and the definition of the problem evolves, causing the specification to alter. There are four possible reasons for these changes:

• Omissions in the original specification
• Changes in the real world
• An increased understanding of the potential benefits of simulation that inevitably occurs as the project progresses
• The identification of new problems through model building and experimentation

The first should not occur if the specification has been communicated properly and feedback has been obtained; however, it must be recognized that the feedback process is unlikely to be perfect.

Whatever the reason for the change, some procedure needs to be put in place to handle it. In the same way that changes to the data are handled (Sec. 7.4) it is useful to freeze the specification once it has been agreed. The project then proceeds in line with this. As a need for change arises it should be assessed for its importance. What effect would excluding the change have on the success of the project? How much would it cost, in time and resource, to implement the change? The costs and the benefits should be understood and a decision made over whether or not to include it. Another possibility is to delay the change and consider it in a second project. If the change is to be excluded, then the reasons need to be justified. If it is to be included, it is important to communicate this clearly, including the impact on timing and resources, and to obtain agreement for the change. Creating a 'change to specification' proforma may be useful for this purpose.

Above all, the aim should be to keep the focus of the project and to ensure that agreed time-scales are met or that changing time-scales are agreed.

Conclusion

Once the project specification has been agreed the problem definition is complete. As the work continues, further changes may be required to this definition, emphasizing the iterative nature of a project. However, for now it is possible to move on to the second phase of the project, model building and testing.

Summary

Why provide a project specification?

- To communicate the project's objectives and approach
- To confirm the validity of the approach
- To ensure credibility
- To reduce the need for changes to the simulation model
- As a reference during and after the project

The structure of the specification

- Introduction to the problem and project objectives
- Expected benefits
- Model summary: scope and level of detail, assumptions, experimental factors, reports
- Data requirements
- Time-scale and milestones
- Estimated cost

Communicating the specification

- To each member of the project team, to the managers, to other interested parties
- Obtain feedback
- Obtain consensus and agreement

- Communicate with: written specifications, formal presentations, verbal specifications

Handling changes to the specification

- Reasons for change: omissions, changes in the real world, increased understanding of simulation's potential, new problems identified through model building and experimentation
- Freeze the specification
- Judge the costs and benefits of changing the specification
- Communicate any changes and gain agreement

Case study 8.1: Panorama Televisions

Example project specification for Panorama Televisions.

Introduction to the problem
Panorama Televisions have been involved in the manufacture of electrical goods since the early days of the radio. They now concentrate on the production of high-quality, premium priced televisions for the international market. There are four televisions in their product range: small, medium, large and flat screen.

Last year, to meet increased demand, Panorama invested in a new television assembly plant. However, the plant has never achieved its target throughput of 500 units per day; in fact, throughput is only just over 400 units.

Panorama propose to investigate this problem with a simulation model. The objectives of this work, the requirements for the model and expected time-scales and costs are outlined in this specification.

Project objectives
The overall aim is:

- To achieve a target throughput of 500 units per day from the television assembly line

The objectives are:

- To determine whether 500 units per day can be achieved with additional pallets only
- To identify the additional storage and pallets required to achieve 500 units per day

The second objective will only receive attention if 500 units cannot be achieved from the first.

Expected benefits
The following benefits are expected from the study:

- Increase throughput by approximately 25 per cent to 500 units per day
- Greater understanding of the television assembly line

Model summary
Scope of the model
The model need only include the television assembly plant specifically from OP10 (load the television to pallet) to OP60 (unload the television from pallet). The final test and pack need not be modelled since it is already known that there is enough capacity in this area.

The following details need to be included in the model:

- General detail: production schedule, number of pallets
- Automatic operations: cycle times, breakdowns, set-ups on changeover, repair/set-up labour
- Manual operations: cycle times
- Conveyors: capacity, transfer times
- Re-work: percentage test failures, re-work times

Assumptions
The following are assumed:

- Conveyor breakdowns are infrequent
- Sub-components, such as television boxes and cathode ray tubes, are 100 per cent available
- Work only takes place during on-shift periods, including repairs to broken machines

Therefore, none of these are modelled.

Experimental factors

- The number of pallets, with an expected range of 30 to 200
- The size of buffers (conveyors), with a maximum 200 per cent increase

Reports

- A time series of daily throughput, both total and for each product
- A histogram of total daily throughput
- The mean, standard deviation, maximum and minimum daily throughput

- Percentage machine utilizations, to include percentage idle, busy, blocked, broken and set-up

Data requirements

DP = data provider

Data required	Responsible	Available by
Physical layout	DP	15 March
Production schedule:		
Batch quantities	DP	16 March
Production sequence	DP	16 March
Number of pallets	DP	15 March
Machines (OP20, OP30, Test, OP50):		
Cycle times	DP	15 March
Breakdowns:		
Time between failure	DP	16 March
Repair time	DP	16 March
Number of repair labour	DP	16 March
Set-ups on changeover:		
Set-up time	DP	16 March
Manual operations (OP10, OP40, OP60):		
Cycle time	DP	16 March
Number of stations at OP40	DP	16 March
Conveyors:		
Capacity	DP	16 March
Transfer times	DP	16 March
Conveyor type	DP	16 March
Re-work:		
Percentage test failures	DP	16 March
Re-work times	DP	16 March
Total number of repair/set-up labour	DP	16 March
Validation data:		
Mean daily throughput	DP	16 March
Distribution of daily throughput	DP	16 March

After 16 March no further changes will be made to the data in the model.

Time-scale and milestones

The work should be completed in 10 working days—by 25 March. The project will be split into four phases:

- Model built   16 March
- Model validated   17 March
- Experimentation complete   24 March
- Final report available   25 March

Estimated cost
It is estimated that a total of 15 man-days of effort are required to complete the project. No training, consultancy, hardware or software costs are necessary.

Case study 8.2: Natland Bank

Example project specification for Natland Bank.

Introduction to the problem
Natland Bank are planning to open a new branch. They are concerned about the facilities required in order to achieve satisfactory service levels, especially with reference to the automatic teller machines. The bank's policy is that customers should not have to wait for more than three minutes to be served. Natland would also like to know whether it is best for customers to wait in individual queues for each service point or in a single queue.

It is proposed to investigate these issues with a simulation model. The objectives of this work, the requirements for the model and expected time-scales and costs are outlined in this specification.

Project objectives
The objectives are:

- To compare the customer service level achieved with one, two and three auto tellers
- To identify which queuing policy gives minimum average waiting times; the choices are a single queue or individual queues for each auto teller
- To convince managers that simulation is a useful tool and the results are valid

If there is time the following objective will also be considered:

- To confirm that the configuration of the other facilities, the manual tellers and the enquiry desk, is correct

Expected benefits

- Identify bank facilities and queue discipline to meet customer service levels
- Effective communication of requirements to management

Model summary
Scope of the model The model need only include the operation of the auto tellers. An investigation of the manual tellers and the enquiry desk will be performed with two separate models, which will not be built until the simulation of the auto tellers is complete, and only if there is time.

The following details need to be included in the model:

- Customers: arrival rates
- Queues: single/multiple queues, queue discipline and jockeying
- Auto tellers: service times

Assumptions It is assumed that auto teller failures are negligible and so thay are not modelled.

Experimental factors

- The number of auto tellers, either one, two or three
- Single queues or multiple queues

Reports

- A histogram of waiting time for each customer, also including the mean, standard deviation, maximum and minimum waiting times
- The percentage of customers served within the three minute service level
- Percentage facility utilization, both in a table and a pie chart
- Cumulative totals showing the mean, maximum and minimum queue sizes and the total customer demand
- A time series showing the queue size and number of customer arrivals every 15 minutes

Data requirements

DP = data provider

Data required	Responsible	Available by
Physical layout	DP	14 November
Customers:		
Arrival rate at auto tellers	DP	14 November
Queue discipline	DP	14 November
Auto tellers:		
Service times	DP	14 November

After 14 November no further changes will be made to the data in the model.

Time-scale and milestones
The work should be complete in five working days—by 17 November. The project will be split into three phases:

- Model built and validated   15 November
- Experimentation complete   16 November
- Final presentation   17 November

Estimated cost
It is estimated that a total of 8 man-days of effort are required to complete the project. No training, consultancy, hardware or software costs are necessary.

PART THREE

Model building and testing

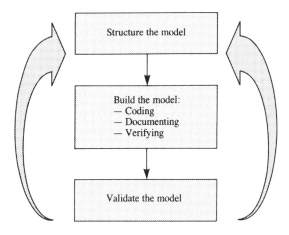

9

Structuring and building the model

The things that can go wrong when a computer becomes involved! On one occasion I was building a model under severe time pressure. I leapt to the keyboard and started typing with little thought as to how best the model could be built. As time progressed, various problems with the model meant that additional sections of code had to be added in a somewhat *ad hoc* manner. The model was eventually littered with various add-ins and little nuances in an attempt to correct the deficiencies of the original structure. Not only was the model code uninterpretable, but time could almost certainly have been saved by planning the structure of the model before attempting to build it. It could be said that I had let the clutch out on my fingers before putting my mind in gear.

Such an experience is all too common. The aim of this chapter is to show how this pitfall can be avoided by structuring the model before building it on the computer. The second aim is to outline the requirements and methods for building, documenting and verifying the model; these are discussed later in the chapter.

Since the actual methods of performing the steps outlined in this chapter are very much dependent on the software package being used, the discussion focuses on the general approach and not on the specifics as to which keys to press. For this reason, the Panorama Televisions and Natland Bank case studies are not given at the end of this chapter. However, if these are being followed, it is recommended that the reader builds and verifies the models before progressing to Chapter 10.

9.1 Structuring the model

As demonstrated above, it is important to structure the model before attempting to build it on a computer. When teaching people to perform simulation projects I always point out that 'a day on paper saves a month on a computer'. Perhaps this is an exaggeration, but the aim is to emphasize the need for proper model structuring away from the computer. Time spent planning is not time wasted but almost certainly time saved. It also ensures that the structure of the model is thought through and that the best modelling methods are considered. Another benefit is that it provides written documentation which can be used throughout the project and whenever the model is used for further work.

Throughout the process of model structuring the aim is to consider how best to create the model described in the project specification (Chapter 8). The intention is not to change the objectives, experimental factors, reports or scope and level of the model, although further deliberation may lead to the conclusion that some changes are required. The primary consideration should be the efficiency of the model, aiming to build the model quickly while also attaining the desired run speed.

The structure should consist of the elements that need to be defined and the data and logic required to drive the model. It is effectively a paper version of the computer model. It may not be possible to define the structure fully and to cover every eventuality at this stage, but at least a broad structure should be created. The actual method of forming the structure on paper depends very much on the particular software that is being used and on the preference of the modeller. Some example methods are:

- Layout diagrams with flows and logic identified
- Activity cycle diagrams
- Flow charts

For some sections of a model it is almost impossible to plan the structure on paper since it involves a particularly complex piece of logic. Rather than attempting to experiment with alternative ideas in the fully fledged simulation, a useful technique is to build a small test model of the section in question. The advantage of this is that test models are contained and run quickly which helps complex logic to be developed fast. In some software packages this approach is enhanced further by enabling these models to be read into a larger model, avoiding the need to retype the code.

9.2 Model building

The process of building the model consists of three distinct activities:

- Coding, entering the model into the computer
- Documenting, explaining the model structure

- Verifying, ensuring the code is correct

This is shown in Fig. 9.1.

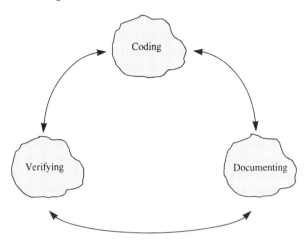

Fig. 9.1   The model building cycle

These activities are best performed in small steps and in an iterative manner, building a section of the model and then documenting and verifying it. Although documentation and verification are vital tasks, for most they hold little fascination. By splitting the model building into smaller steps and performing these tasks at each stage, lengthy periods of documentation and verification are avoided which can only serve to reduce the tedium in performing them.

The actual sequence in which documentation and verification are performed depends upon the preference of the modeller. This is shown by the two-way arrows in Fig. 9.1.

The activities of coding, documenting and verifying are now discussed in detail.

MODEL CODING

The requirements of the simulation package basically dictate the method for entering the model into the computer. It is recommended that this process is split into a number of small steps. Figure 9.2 shows an outline approach to incremental model coding. Here a small section of the model is coded and the level of detail is gradually increased. Once one section of the model is complete, including documentation and verification, the modeller moves on to the next. The number of sections depends upon the structure of the model and in some cases it is most convenient to consider the whole model as one section.

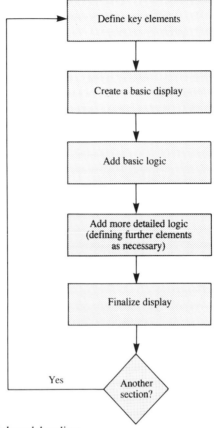

Fig. 9.2   Incremental model coding

This approach needs to be adapted when there is model logic which crosses section boundaries, for example a scheduling system or control policy. When these need to be modelled it is probably best to code the simulation in an incremental manner as far as possible and then to add the 'macro' logic at a later stage.

Throughout the process of model coding it is important to aim for:

- Speed of model coding
- Ease of understanding
- Fast run speed

This is likely to amount to some compromise on all three. For any portion of model code there are probably at least a dozen ways of creating it. On many occasions I have discussed methods of coding with colleagues who have thought of completely different approaches to my own. The important thing is to consider alternative methods and to select the one that is most efficient, that is the one which gives the best compromise on these three factors.

Use of pseudo-random number streams (PRNS)

An area of particular importance while creating the model code is to make proper use of pseudo-random number streams (PRNS). Most simulation packages ask for a PRNS whenever there is some form of random behaviour that is normally expressed through the use of distributions (Sec. 7.5). The question arises as to how the PRNS should be selected. It is recommended that a PRNS should only be referenced once in a model. In other words, when the first distribution is entered into the code PRNS 1 is used, when the second distribution is included PRNS 2 is selected and so on. This takes advantage of the fact that a PRNS generates the same values in every experimental run and therefore when comparing alternative scenarios it is possible to make direct comparisons.

To illustrate this, take a simulation in which there are three machines working in parallel (Fig. 9.3). As the model progresses so the machines break down in sequence (an assumption that is made for the purposes of simplicity) and the repair time is sampled from a uniform distribution with a minimum of five minutes and a maximum of fifteen. If the rule above is broken and PRNS 1 is used for all the machines the repair times given in Fig. 9.3 would be sampled.

Having performed this experiment, a second run is performed with only two machines (Fig. 9.4). The repair times given in Fig. 9.4 would be sampled.

If a comparison of the throughput from the two experiments were made then two factors have changed: firstly, the number of machines and, secondly, the repair times. It is therefore difficult to draw conclusions regarding the relative benefits of the two systems.

However, if different PRNS had been used to sample the repair times of each machine, the breakdowns would have been the same in the first and second experiments, enabling a comparison of like with like. This is technically known as variance reduction using the method of common random numbers (Sec. 11.5).

MODEL DOCUMENTATION

The purpose of the documentation is to provide a record of the simulation that the model builder or another modeller will be able to pick up and use at some future date in order to make changes to the model. It should also

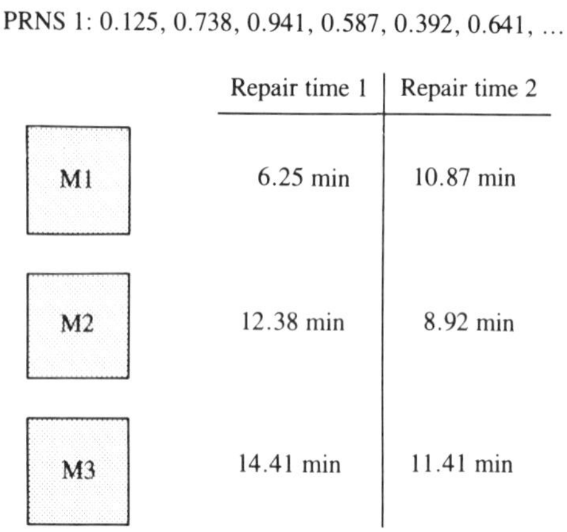

PRNS 1: 0.125, 0.738, 0.941, 0.587, 0.392, 0.641, ...

	Repair time 1	Repair time 2
M1	6.25 min	10.87 min
M2	12.38 min	8.92 min
M3	14.41 min	11.41 min

Fig. 9.3 Single PRNS used across three machines

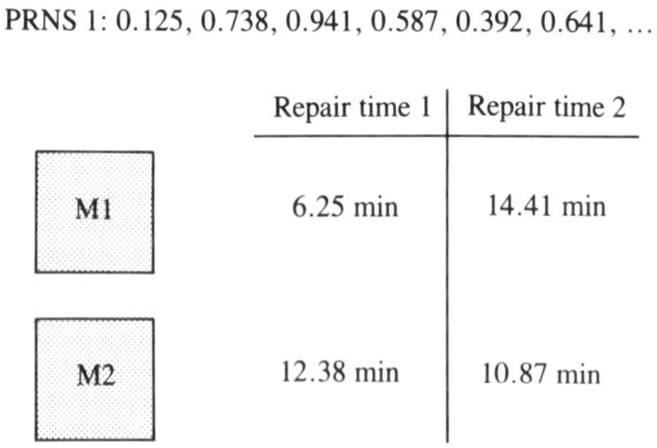

PRNS 1: 0.125, 0.738, 0.941, 0.587, 0.392, 0.641, ...

	Repair time 1	Repair time 2
M1	6.25 min	14.41 min
M2	12.38 min	10.87 min

Fig. 9.4 Single PRNS used across two machines

help in providing information to the modeller during the current project. Specifically the documentation should provide:

- A list of elements and variables, and their purpose
- A summary of the model data
- An explanation of the model logic

The model structure (Sec. 9.1) already provides some of this information. This must be updated while the model is being coded to include any alterations that are made. For variables and elements it is sensible to use meaningful names and avoid the temptation to employ algebraic symbols such as X, Y, A and B; for example, using a variable named OP10CYC for the cycle time of OP10 can only aid understanding. Most simulation packages provide some documentation features; text files of the model code, summaries of element details and reports on cross-referenced elements and variables are often provided in the software. Facilities for adding additional comments should be utilized to the full in order to explain the purpose and structure of the code. The provision of written and diagrammatic explanations on paper are also beneficial.

It is vital that the documentation is written during the coding of the model, otherwise the modeller will soon have forgotten what has been done and why. Also, for most people documentation could hardly be described as a desirable task and leaving it until the end means that the work will only build up, reducing the chances of it being done well, if at all.

Having created the documentation, all that remains is to collate the information into some logical sequence, the aim being to have a full set of documentation, both written and software based, which can be used during and after the project. It should be noted that the project specification (Chapter 8) provides useful information in addition to this documentation.

Documentation for the model user

In Sec. 2.5 an outline of a simulation project team is given and the potential for a model user to carry out experimentation is discussed. In this situation the general model documentation described above needs to be supplemented by additional information. In fact, the documentation may be of little use to a model user since they probably have no expertise in model building. What is required is user documentation that describes how experiments can be performed with the model. As a guide, the content of such documentation should include:

- Objectives of the project
- Overview of the model
- Experimental factors, description, range and how to change them
- How to run the model and perform experiments
- Results, how to access, analyse and interpret them

Much of this user documentation can be taken from the project specification (Chapter 8) and the general model documentation. Therefore, it is probably best written once the model has been built and is known to be valid (Chapter 10).

MODEL VERIFICATION

In the simulation literature there are a number of alternative views regarding the purpose and definition of verification and validation (Chapter 10). Here specific definitions are used and in Sec. 10.1 the differences between verification and validation are highlighted.

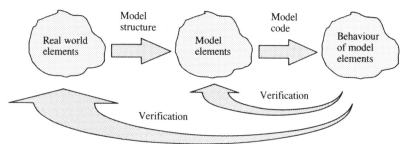

Fig. 9.5   Model verification

The purpose of verification is to guarantee the correct behaviour of each element in the model. For example, in a model of a supermarket the behaviour of the check-outs needs to be verified, ensuring that the service times and staff rosters are modelled correctly. Verification is effectively a 'micro' check of the model. (For those with a background in computer programming, model verification is analogous to program debugging.)

Figure 9.5 summarizes the process of verification. Creating a simulation model requires two distinct steps: firstly, the model structure is planned and, secondly, the model is coded using a simulation package. In this process the elements in the real world are transformed into model elements. The model is then coded and the simulation run, resulting in specific behaviours for each element in the model. During verification this process is reversed. Each element is tested in turn to ensure that, firstly, they behave in the manner intended by the model code and, secondly, that their behaviour is representative of the real world. For example, the modeller might have been informed that the cycle time of a machine is 2.6 minutes. This has been included in the model structure and coded. During verification the modeller must check that, as the simulation runs, a cycle of 2.6 minutes is actually achieved and also that this behaviour is correct when compared to the real world. For the latter it is important to have regular contact with experts in the system being modelled. These experts can then verify that the elements in the model are behaving correctly.

A number of specific areas should be tested during verification. These include:

- Timings, e.g. cycle times, repair times, travel times
- Control of elements, e.g. breakdown frequency, shift patterns
- Control of flows, e.g. routing
- Control logic, e.g. scheduling rules, stock replenishment
- Distribution sampling, e.g. gamma input data gives gamma samples

Normally it is not necessary to verify all of these since some appear as standard features in many simulation packages, for instance distribution sampling. However, software packages do contain bugs and it is not always safe to assume that the data are being handled correctly. Another problem is that the software may interpret some data in a different way to that which the modeller intended; a common example is the calculation of the time between failures. If there is any doubt, the model behaviour should be checked during verification.

It is worth repeating here that verification should be performed while the model is being coded. In this way it is possible to ensure that each element in the model is checked as the logic and data are being input into the computer.

The specific methods of verification can be split into three broad categories:

- Checking the code
- Visual checks
- Inspecting output reports

None of these should be used exclusively, but in conjunction with each other. Each of these methods are now discussed.

Checking the code

The modeller needs to read through the code to ensure that the right data and logic have been entered. This is especially true for areas of complex logic. A useful idea is to get someone else to read the code as a second check. If no modelling experts are available then most simulation software vendors offer a help desk service with which specific areas of code could be discussed. Alternatively, by expressing the code in a non-technical format (the documentation could be used for this purpose) a non-expert could verify the data and the logic.

When I have been unsure about a section of complex code I have found it useful to sit down and go through it with someone else. The mere discipline of trying to explain my work to another person has meant that I have been able to spot errors and rectify them, very often without actually needing any intervention from my colleague.

Visual checks

The visual display of the model proves a powerful aid in model verification. By running the model and watching how each element behaves, both the logic of the model and the behaviour against the real world can be checked. Various ideas aid this approach:

- Stepping through the model event by event
- Stopping the model, predicting what will happen next, running the model on and checking what happens
- Interactively setting up conditions to force certain events to take place
- Isolating areas of the model so it runs faster, reducing the time to perform thorough tests
- Explaining the model to facility experts in order to gain their opinion
- Tracing the progress of an item through the model

Inspecting output reports

By inspecting the reports from a simulation run, the actual and expected results can be compared. Of particular interest at this stage is the performance of individual elements, for example their utilization or work-in-progress. Graphical reports of samples from input distributions, for instance machine repair times, are an aid in checking that they are being modelled correctly.

A report that may be of some use is a 'trace' of a simulation run. This is a blow-by-blow history, written to a text file, of every event that took place during the run. Inspecting this report can help to diagnose and rectify problems.

Conclusion

In this chapter the need to structure the model before building it on a computer has been emphasized. The model is then built in three iterative steps: coding, documenting and verifying. All are vital for a successful simulation. Following this step, the model needs to be validated.

Summary

Structuring the model

- A day on paper saves a month on a computer
- Structure consists of elements, data and logic
- Use test models to create complex logic

Model building

- Small steps iterating through: coding, documenting, verifying
- Model coding
  –Split into small steps
  –Incremental increase in depth of logic
  –Aim for: speed of model coding, ease of understanding, fast run speed

–A pseudo-random number stream should only be referenced once in a model
- Model documentation
  –Provide documentation on: elements and variables and their purpose, a summary of the model data, an explanation of the model logic
  –Documentation tools: model structure, meaningful element names, software documentation features, written and diagrammatic explanations
  –Provide documentation for a model user if it is required
- Model verification
  –Check that each element in the model behaves correctly
  –Areas to check: timings, control of elements, control of flows, control logic, distribution sampling
  –Methods of verification: checking the code, visual checks, inspecting output reports

10

Model validation

The excitement of having a working model can prove all too much. Pressed by a tight time-scale and keen to obtain results immediately the project team can very easily move straight from model building to experimentation. The results will certainly be available more quickly but can they be trusted? Before the first experiment is run it is vital that the model is fully tested to ensure that it is accurate and that it can meet the objectives of the project. Any errors that are discovered can then be corrected. Only when these tests have been completed can experimentation be performed with confidence. This aspect of the project is known as model validation.

The aim of this chapter is to show how a model can be validated. First there is a general discussion on the purpose of validation, how it differs from verification and the issue of credibility. A number of validation methods are then outlined as well as the need to ensure that the general project objectives have been met. Finally, there is a discussion on the need to analyse the sensitivity of the results to any unobtainable data that have been estimated. The chapter concludes by applying some of these ideas to the Panorama Televisions and Natland Bank examples.

10.1   Validity and credibility

As with verification there are a number of alternative views in the simulation literature regarding the purpose and definition of model validation. Here one specific definition is used. A valid model is both accurate and able to meet the objectives of the simulation project for which it is being used. The purpose of validation is to guarantee the correct degree of accuracy by checking that the overall behaviour of the model is representative of the real world. The accuracy of the model need only be sufficient to meet the objectives of the project (Sec. 6.2). Model validation can be seen as a 'macro' check of the simulation. It is important to note that a model is only valid for

Table 10.1   Model verification and validation

Verification	Validation
'Micro' check	'Macro' check
Performed during model coding	Performed on completion of the model
Test each element individually	Test the overall model accuracy and its ability to meet the project's objectives

the purpose for which it is built. If at some point the model is to be used for another purpose then its validity within this context must be checked first.

Credibility, as well as validity, is an important consideration at this stage. This need is discussed in Chapter 8 as one of the reasons for providing a project specification. It is now important to obtain credibility for the working model. Without this, although the results may be very accurate, the project will fail through lack of confidence. The methods of validation discussed below should also act as an aid in gaining credibility.

VERIFICATION VERSUS VALIDATION

During verification (Chapter 9), an investigation of each element in the model is performed. This is a precursor to validation; if each element is not verified as correct the model is extremely unlikely to be valid. However, validation goes a step further by considering the accuracy of the overall model.

Although verification and validation are described as separate steps there is a significant amount of overlap between them. Some of the methods employed are very similar and it should be no surprise when verification continues during the validation phase. I was involved with a project in which an automatic guided vehicle system was being modelled. During validation a number of errors in the vehicle control logic were identified as a result of inaccuracies in the overall model behaviour. These were traced back to their cause, rectified and verified before model validation continued. The differences between verification and validation are summarized in Table 10.1.

10.2   Methods of validation

A number of validation methods are now discussed. These are grouped into three specific types:

- Face validity

- Comparison with the real system
- Comparison with other models

The choice of method depends upon its appropriateness and the preferences of the project team. However, no method should be used in isolation and it is recommended that a number of validation checks are performed. If the simulation is found to be invalid then the reasons for this should be investigated, the model updated and its validity reassessed. Throughout this process the purpose is to ensure that the model is suitably accurate and that the project's objectives can be met.

A lack of real world data for comparison causes some difficulty when trying to validate a model. This is obviously a problem when a new facility is being designed but data may also not be available for an existing facility. In both cases the methods outlined below should still be employed; however, it is necessary to rely upon the experience and intuition of someone who is knowledgeable either about the real or proposed system. When doing this it is best to ask the expert to predict the outcome of a simulation before performing a run and then to compare the results. In this way the expert's judgement is not biased by viewing the results before predicting the outcome. If the expert's prediction and the model's result are at variance then this difference should be explained; either the model is not valid, the prediction incorrect or both.

For some of the validation methods outlined below an experimental run is required. In such cases the model must be warmed up and then a number of short runs or a single long run performed. These issues are discussed in detail in Chapter 11.

FACE VALIDITY

A simulation is said to have face validity when, on the surface, it appears to be reasonable to those who are knowledgeable about the system being modelled. Two methods for obtaining face validity are discussed here.

'Watch the model run for an hour'

Very often I have sat for an hour and watched a model as it runs. There is a lot to be learnt by simply watching a simulation; in fact, it is surprising how often people do not take full advantage of the visual display in this way. As the model runs it is good to be active and ask questions such as:

- What do I expect to happen next?
- Why is that machine always blocked/idle?
- Why is a large queue building up?

Sometimes it is useful to interact with the model and force certain conditions in order to perform validation under specific circumstances. This process is different from the visual checks for verification in that the performance of the model as a whole is now being considered.

Demonstrate the model

An extension of the idea above is to demonstrate the model to the simulation project team, the managers and any other interested parties. During the presentation the model can be run and the range of inputs and reports presented. Feedback should be obtained on the accuracy of the model and its ability to meet the project's objectives. Allowing the model to be scrutinized in this way not only enhances the validation process but also provides a useful means for increasing the credibility of the work. For this reason this method of validation should be employed whenever possible.

COMPARISON WITH THE REAL SYSTEM

During validation it is particularly important to compare the performance of the model against the real system. Data should be collected for this purpose (Sec. 7.1). If the real system is not in existence then comparison can be made to expectations and intuition, possibly gained from the requirements of the design or from experience with similar systems. Three specific methods are now discussed.

Historic data

Historic data collected from the real system such as throughput, work-in-progress and customer service levels can be compared to the results of the simulation. It is important to check not only the average levels of these data but also to compare their spread. This can be performed by judging how closely the averages from the model and the real world match, and by visually comparing the distributions of the data. For those with a knowledge of statistics the distributions can be compared using statistical techniques (Sec. 14.2).

Input/output relationships

The experimental factors and possibly model data are changed, the model is run and results obtained. The relationships between the inputs (experimental factors and model data) and the outputs (results) should be the same for the model as for the real system. For example, changing the size of a storage area should have a similar effect on the model as it does in the real world.

Turing tests

In a Turing test the model reports are made to look exactly the same as the reports provided by the real system. One or more reports from the model and from the real world are given to someone who is knowledgeable about the system. They are then asked to try and distinguish between the two. If they are unable to detect any difference this is indicative of model validity. Even if real world reports are not available it is still worth asking an expert to review the model reports.

COMPARISON WITH OTHER MODELS

The final group of validation methods is a comparison of the simulation with other models. These are particularly useful when the real system does not exist. Three methods are outlined.

Comparison with mathematical models

It is unlikely that a mathematical model is able to predict the outcome of the simulation exactly, otherwise the simulation would probably not have been built in the first place. However, by simplifying the simulation it may be possible to draw a direct comparison with a mathematical model, for example paper calculations, a spreadsheet analysis or queuing theory.

Deterministic models

An extreme version of the above concept is to simplify the model to the degree that there are no random events, in other words it becomes a deterministic model. It should then be possible, in most cases, to predict the outcome of a simulation run exactly. This is a very useful means for testing not only the validity of a model but also for obtaining a greater understanding of the facility being modelled. In one project I performed a test with a deterministic model and found that throughput was nearly 25 per cent below expectation. On further investigation it was found that there was a serious design fault in the proposed system which was duly rectified. However, had such a test not been performed, the shortfall in throughput may have been attributed solely to machine breakdowns and the design fault never found.

Comparison with similar simulation models

If a similar model exists then the results of the two simulations could be compared. This method should be used with caution since it is obviously necessary to be sure that the similar model is valid before any comparisons are performed.

10.3   Validating the general project objectives

In Sec. 4.4 a number of general project objectives are outlined. These are:

- Time-scale objectives
- Run-speed objectives
- Visual objectives
- Interactive objectives

The time-scale objective should be checked as part of the on-going management of the project and does not need specific consideration here. However, during validation it is important to ensure that the other requirements are met. It is not unusual for these to have changed during the project and some re-modelling work may therefore be necessary.

RUN-SPEED OBJECTIVES
A target run speed may have been set as part of the project definition. Has this been achieved? If not a number of options are available:

- Consider more efficient modelling methods
- Attempt to simplify the model
- Obtain faster hardware
- Accept the slower run speed

VISUAL OBJECTIVES
Is the display adequate for the purposes of the project? Conformance with the visual objectives can be checked with the project team, especially the client, and any necessary changes made. Care must be taken to contain these changes within the scope of the project's objectives in order to avoid unnecessary additional work.

INTERACTIVE OBJECTIVES
Is the level of interaction correct for the needs of the project? Again, conformance can be checked with the project team and the model updated accordingly.

10.4   Testing the sensitivity to unobtainable (category C) data

In Sec. 7.3 methods of dealing with unobtainable (category C) data are discussed. In general these data need to be estimated. During validation it is important to determine the effect of any inaccuracies in these data by

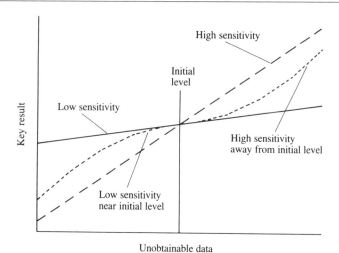

Fig. 10.1   Sensitivity analysis on unobtainable (Category C) data

performing some sensitivity analysis. If the results of the simulation are sensitive to these data then further action needs to be taken.

In sensitivity analysis the values of the estimated data are changed and the model run to see the effect on the key results, for example changing the level of breakdowns on machines to discover the effect on throughput. It is probably best to start by making significant changes, for instance doubling and halving values, in order to obtain an initial feel for the sensitivity; these limits could be estimated by considering the likely extremes of the data. After this, smaller incremental changes could be made. It is recommended that a minimum of three values are used (the initial level, a minimum and a maximum) since the response of the key results may not be linear. If time allows, it is best to perform further analysis. In each case the model must be warmed up and run (for a discussion on these see Chapter 11).

Having performed the sensitivity analysis a graph could be drawn such as the one shown in Fig. 10.1. If the results prove to be sensitive to the data then two options are available. Firstly, some effort could be made to improve the information; however, since the data are in category C this is probably not possible. Secondly, further sensitivity analysis could be performed during experimentation and when the results are presented an expected range of outcomes reported. Obviously, if the results are not sensitive no further action is required.

If the sensitivity analysis is to be performed on more than one set of data, the interaction of these data must also be investigated. By performing individual analyses on the data quite different conclusions may be drawn

Time between failure

		−25%	Initial level	+25%
	−25%	79	105	124
Repair time	Initial level	85	112	131
	+25%	92	118	135

Results on daily throughput

Fig. 10.2 Sensitivity analysis on two sets of unobtainable (Category C) data

than by performing the analyses in tandem. Therefore, the sensitivity analysis needs to be adapted. All the sets of data should be changed both independently and together in order to obtain a feel for their individual and combined effect on the simulation result. The results of this analysis could be summarized in a table such as that shown in Fig. 10.2. Here the time between failures and repair times of machines have been varied in order to judge their effect on throughput. By inspecting this table it should be possible to judge the level of sensitivity. In this example it would appear that throughput is more sensitive to the time between failures than repair times.

Obviously, such an analysis may require a large number of simulation experiments. These can be reduced by selecting a limited number of runs and using the results to judge the likely outcomes of the other experiments (see the second part of Sec. 12.1).

Conclusion
The model has now been validated and any errors should have been addressed. The sensitivity of the simulation results to any unobtainable data that have been estimated is also known. It is now possible to proceed with experimentation being confident in the validity of the results.

Summary
Validity and credibility

- Validation 'macro' check, performed on completion of the model, tests the overall accuracy of the model and its ability to meet the project's objectives
- credibility: required for confidence in the results

Methods of validation

- Employ a number of validation checks
- Investigate and correct errors
- When no real system for comparison, use experience and intuition
- Face validity: 'watch the model run for an hour', demonstrate the model
- Comparison with the real system: historic data, input/output relationships, Turing tests
- Comparison with other models: comparison with mathematical models, deterministic models, comparison with similar simulation models

Validating the general project objectives

- Check that the run-speed objectives have been met
- Check that the visual objectives have been met
- Check that the interactive objectives have been met

Testing the sensitivity to unobtainable (category C) data

- Determine the effect of any inaccuracies in the data
- Perform sensitivity analysis
- If the data prove sensitive: try to improve the data, investigate the sensitivity of the experimental results
- Adapt the sensitivity analysis for two or more sets of data

Case study 10.1: Panorama Televisions

Three validation checks have been performed on the Panorama Televisions model.

Watch the model run for an hour
The model was watched for a period of running time to ensure that the overall behaviour appears reasonable.

Deterministic model
All stochastic elements of the model were removed, that is:

- Breakdowns
- Set-ups
- Television repairs
- Variation in cycle times

The model was then warmed up for a period of one hour (Sec. 11.3) before collecting results over a further 24 hour period.

The expected throughput can be calculated from the cycle of the slowest machine which is 2.1 minutes. Therefore:

$$\text{Expected throughput} = \frac{60 \text{ minutes} \times 24 \text{ hours}}{2.1} = 686 \text{ units}$$

The actual model throughput was 685. The difference can be attributed to rounding errors.

Historic data
Historic data on throughput over 50 days has been provided (Appendix 2). Following a warm-up period of 12 hours (Sec. 11.3), results were collected from the model for a similar period of 50 days. The mean daily throughput from the model and the real world were:

	Model	*Real world*
Mean daily throughput	413	407

The distributions of daily throughput are compared in Fig. 10.3. The results appear to match favourably.

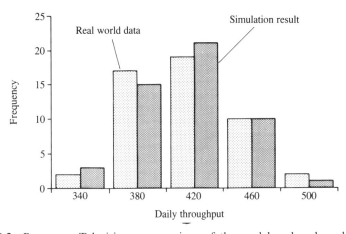

Fig. 10.3 Panorama Televisions: comparison of the model and real world daily throughput

Case study 10.2: Natland Bank

No historic data are available for the Natland Bank model since it is a new facility. However, three validation checks have been performed on the model.

Watch the model run for an hour

The model was watched for a period of running time to ensure that the overall behaviour appears reasonable.

Deterministic model

All stochastic elements of the model were removed, that is:

- The arrival rate was fixed at 100 per hour
- The service time was fixed at 1 minute
- A single queue was specified

The model was then run for one day of operating time, that is six hours, with both one and two tellers.

The expected queue length at the end of a day's business and the teller utilization can be calculated as follows:

$$\text{Expected queue length} = \text{hours} \times (\text{arrivals per hour} - \text{customers served per hour})$$

$$\text{Expected teller utilization} = \text{minimum} \begin{cases} 100.0 \\[2ex] \dfrac{\text{arrivals per hour}}{\text{customer served per hour}} \\[2ex] \times 100.0 \end{cases}$$

The data for the model are:

Arrivals per hour $= 100$
Customers served per hour with one teller $= 60$
Customers served per hour with two tellers $= 120$

Therefore at the end of six hours:

	One teller	*Two tellers*
Expected queue length	240	0
Expected teller utilization	100%	83.3%

From the model the following results were obtained:

	One teller	*Two tellers*
Queue length	240	0
Teller utilization	100%	83.3%

An exact match between the model and the expected result has been achieved.

Comparison with mathematical models
The results of the full model have been compared to expectations based on average arrival rates and service times. Some discrepancy was expected since the random variation in the model cannot be accounted for in a static calculation. Three experiments have been used for comparison:

- 1 teller, 1 queue
- 2 tellers, 2 queues
- 3 tellers, 3 queues

The mean service time and arrival rates have been calculated from the data in Appendix 2. These calculations are not included here but their results are:

Mean arrivals per hour   = 146.7
Mean customers served per hour   =   96.9 per teller

Therefore, the expected queue length at the end of a day's business and the teller utilizations have been calculated in the same manner as for the deterministic model:

	One teller	*Two tellers*	*Three tellers*
Expected queue length	299	0	0
Expected teller utilization	100.0%	75.7%	50.5%

The model was then run and the following results obtained:

	One teller	*Two tellers*	*Three tellers*
Queue length	343	11	1
Teller utilization	98.2%	80.2%	54.0%

There is some variance between the calculated results and those of the simulation. This can be attributed to the variation in customer arrival rates throughout the day. In general the match appears adequate.

Note that in the above experiments six runs of the same model were performed with different pseudo-random number streams. The results reported are mean averages for the six runs. The need for these multiple runs (replications) is discussed further in Sec. 11.4.

PART FOUR

Experimentation

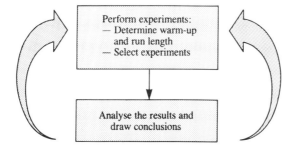

11

Experimentation

The results of the simulation were showing a 10 per cent shortfall in throughput. It was a large model which ran slowly and so it had been decided that experiments should be run for only one week of simulated time. Some effort had gone into achieving the target throughput by changing various experimental factors and re-running the model. However, little progress had been made. Eventually, someone questioned whether the first week of production was representative of normal working practice. At this point the decision was taken to leave the model running for some time in order to find out what would take place in subsequent weeks. Having left the model running for a long weekend results were obtained for a number of weeks of production. On analysis it was found that the first week proved to be particularly poor and that for the majority of the subsequent weeks throughput was much improved. In fact, average performance was not far short of target.

It is the purpose of this chapter to show how this kind of situation can be avoided by employing the correct experimental procedures. Models must be warmed up and run for a suitable time in order to obtain reasonable results. There are various decisions that have to be made:

- What length of warm-up period is required?
- Can starting conditions be used?
- How many replications are required?
- How long should a long run be?

There are no absolutes in answering these questions, but there are methods that can be employed to give reasonable answers. These methods are discussed in this chapter. Before addressing these issues, there is an overview of the different types of simulation model and experiment. These need to be identified in order to determine which experimental procedure should be employed. The chapter concludes by applying these principles to the Panorama Televisions and Natland Bank case studies.

11.1 Types of simulation model

Before the specifics of experimentation are considered it is important to understand the various types of simulation model and their associated characteristics; in this way the correct experimental methods can be applied. Three characteristics are now discussed:

- Deterministic and stochastic
- Steady-state and transience
- Terminating and non-terminating

DETERMINISTIC AND STOCHASTIC MODELS

A deterministic simulation does not contain any random events; all the data in the model are set at predetermined levels. A stochastic model has one or more random events; for example the cycle times, failures and arrival rates are sampled from statistical distributions. A model is stochastic if distributions and pseudo-random number streams are referenced in the model code, otherwise it is deterministic.

Obviously the results of a stochastic model are subject to variability, for instance, the throughput of a manufacturing facility changes from day to day. However, it does not necessarily follow that the results of a deterministic model are constant; the complexity of the model may create some form of unpredictable or pseudo-stochastic behaviour. In Sec. 7.5 there is a discussion on the use of data streams for modelling random events. Such a model could be described as deterministic. However, if the results were recorded on, say, a daily basis, they would probably be subject to variation and appear to behave in a stochastic manner.

STEADY-STATE AND TRANSIENCE

A simulation is said to be steady-state if the probability of being in any particular condition is always the same; for example there is always a 10 per cent chance that there are five customers in a queue. Steady-state should not be confused with equilibrium where the condition of the simulation does not change during a run; in steady-state the condition of the simulation varies as time progresses. An example of a steady-state simulation is a model of a manufacturing facility in which the throughput varies from day to day; however, the data that drive the simulation are always being sampled from the same distributions.

A particular instance of steady-state is a steady-state cycle simulation. Here the probability of being in any particular condition varies within a 'cycle' of an experimental run. A cycle is the period over which the distributions of the data that drive the simulation vary. For example, the size of a queue at a petrol station forecourt varies, not only because of the random interarrival time of the cars but also because the average arrival rate changes during the day; in other words, the distribution of the arrival rate changes. The cycle is complete when the end of the sequence is reached, in this example after a day. In a model that tests a monthly production schedule, for instance, the cycle length is one month. A model can be identified as a steady-state cycle if the data in the model changes according to some set sequence as a run progresses.

Transience occurs when a simulation is not in steady-state. This takes place during the warm-up period (Sec. 11.3) before the model has reached a normal working condition. However, in some instances experiments are purposefully performed during, or by creating, a transient period, for example testing the effect of a major catastrophe on a production facility. Most simulation models in the Operations Management arena are steady-state unless it has been specifically decided that transient conditions should be explored.

TERMINATING AND NON-TERMINATING SIMULATIONS
A terminating simulation is one for which there is a natural end point to an experimental run. This might be, for example, the end of a day's business in a supermarket or the completion of a production schedule in a manufacturing facility. For a non-terminating simulation there is no such end point and the model could theoretically be run for an infinite period of time.

11.2   Methods of experimentation
There are broadly two types of experimentation.

INTERACTIVE EXPERIMENTATION

Interactive experiments are performed by running the model, watching what happens, deciding on actions and implementing them in the model to see their effect. For instance, by watching a model run it might be noted that there is a build-up of work-in-progress which is caused by poor utilization of resources. Corrective actions could be identified which are then implemented in the simulation and the run continued to see their effect.

Performing experiments in this way can prove a very useful technique, especially as an aid to group decision making, for investigating specific conditions and for training staff. It also enables ideas to be filtered, creating a short-list of potential solutions, and thereby acts as a useful precursor to more thorough experimentation.

Care must be taken when performing interactive experiments. Since the model is only viewed for a relatively short period of operation the results obtained are unlikely to represent the full range of operating conditions. Therefore, more thorough experimentation should always follow interactive runs in order to confirm, or deny, the validity of the results.

BATCH EXPERIMENTATION

Batch experiments are performed by setting the experimental factors and leaving the model to run for a period of time in order to obtain some results. Most simulation software allows the display to be switched off and the model to be run forward to a specific point in time at which the display is refreshed to show the current state of the model. Often a number of experiments can also be set up to run one after the other without the need for further intervention from the model user. Batching forward in this way improves the run speed of the model and enables a set of experiments to be performed overnight.

The purpose of the experiment needs to be considered. Is the aim to obtain results that are comparative or predictive? Comparative experiments are used to contrast the likely outcomes of two or more alternative scenarios and decide which is best, for instance to show that employing three maintenance operators is an improvement over employing two. Predictive experiments aim to give an absolute value on performance, for example a prediction of the customer service level achieved by a telephone enquiry centre. Since precision is of greater importance, predictive experiments need to be performed more rigorously than comparative ones.

Specific decisions need to be made regarding the design of the experiments. These are now discussed throughout the rest of this chapter and also in Chapter 12. The discussion largely centres around batch experimentation, although some of the ideas could also be applied to the interactive approach.

11.3 Warm-up and starting conditions

When an automotive assembly plant is being brought into service some time is required to get the plant up to speed. The production director is unlikely to expect the first week's throughput to be anywhere near representative of the system's capability. Indeed, some months will probably have passed before the plant reaches capacity. In a similar way the results obtained at the beginning of a simulation run are unlikely to be representative. The model probably starts from empty with no parts, customers, work-in-progress or resources available, and therefore the results reflect this situation. Having run the model for a period of time it becomes more representative of normal working practice. To use more formal language, the model passes through an initial transient period before reaching steady-state.

For the results of a simulation to be correct, they should only be collected once steady-state has been reached. If not, the results are likely to contain bias from the transient condition of the model at the beginning of the run. There are three ways of dealing with this:

- A warm-up period
- Starting conditions
- Mixed starting conditions and warm-up

Each of these are now discussed. The principles are demonstrated by the Panorama Televisions and Natland Bank case studies at the end of the chapter.

WARM-UP PERIOD
When using a warm-up period the model is run until steady-state has been reached and then the collection of results commences. The question arises: how can the warm-up period be determined? A practical 'rule-of-thumb' is outlined here while a statistical procedure is discussed in See 14.3.

Choosing a warm-up period

A practical way of deciding whether a model has warmed up is to ask the question:

- Does the state of the model represent normal working conditions (steady-state)?

This can be answered by inspecting a number of relevant reports, typically:

- Throughput
- Work-in-progress
- Queue/buffer sizes

Also, by viewing the model's display, conclusions can be drawn regarding the state of the simulation.

Using an example of a manufacturing plant, a time-series of throughput and work-in-progress could be drawn. In order to determine the warm-up period it is important to monitor these values closely; therefore, the time interval between observations should be shorter than would normally be used for the full results, say one hour. Figure 11.1 shows an example of the

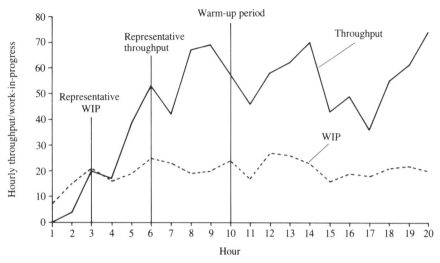

Fig. 11.1 Time-series of throughput and work-in-progress for a manufacturing simulation

two time-series. By inspecting these time-series the points at which the throughput and work-in-progress first represent normal conditions can be located; these points are shown in Fig 11.1. Also, by viewing the model as it runs, consideration can be given to the spread of work-in-progress and the working of the machines, buffering, transportation and resources. Are these performing in a realistic fashion?

The warm-up period can be determined by the point at which the model is first seen to be in a realistic condition. In Figure 11.1 a warm-up period of 10 hours has been selected since it seems reasonable to allow the model to run for some time in a representative (steady-state) condition before collecting any results.

Practical considerations

A number of practical considerations need to be discussed. Firstly, it is best for the warm-up period to be too long rather than too short; otherwise the results will include some bias from the initial transient period. Be generous in your estimates of warm-up time. The disadvantage of this approach is the overhead in longer experimentation times caused by a longer warm-up period.

Most simulation software enables the user to run the model for a period and then to save its status. At a later date, this save can be restored and the run continued. By saving the model at the end of the warm-up time the overhead of re-running the same period can be avoided. However, there is a limitation to this approach. If an experimental factor is only changed at the end of the warm-up period it may be some time before this change takes effect in the model. For instance, it is likely that the times of the next machine failures have already been sampled and a change to these experimental factors will only take effect after the next breakdown. In this situation, either the model needs to be run until the changes have taken effect (a secondary warm-up period) or the whole warm-up period should be run again with the changes to the experimental factors included from the beginning. Careful consideration should be given before saves of the model status are used in place of running the warm-up period.

The warm-up period should be determined separately for each experiment. However, since time-scales are often short and in order to maintain some sense of compatibility between runs, the same warm-up period is often used for every experiment. If this is the case, it is recommended that the warm-up period is regularly checked to ensure that it is correct as well as checking it for any runs from which conclusions are drawn. Since the same warm-up period is being used for every experiment, it is also recommended that the warm-up time is overestimated to minimize the possibility of bias from the initial transient period in any of the runs.

Warm-up periods are required for all models, including stochastic, deterministic, terminating and non-terminating simulation, that is unless starting conditions have been defined. In the process of choosing the warm-up period it may be found that the model never reaches steady-state, for example a queue continues to increase in size. This should be viewed as a simulation result and solutions to the problem sought.

As a final point it should be noted that most simulation packages provide facilities for setting the warm-up period. The time can be input to the model code and the results collection automatically commences from the end of the warm-up period.

STARTING CONDITIONS

As an alternative to a warm-up period starting conditions can be used; the simulation is put into a realistic state at the beginning of the run. Materials, resources, work-in-progress and customers are all placed into the model in an attempt to create realistic conditions. This information could be read directly from a facility control system or database. The advantage of this approach is that no time is spent waiting for the model to warm up during experimentation. It also enables complete control over the exact conditions of the model. The disadvantage is that it may not be easy to obtain information on a realistic state for the model, especially if it is a new facility, and it may not be possible to implement the starting conditions exactly due to limitations in the simulation software. When a model is being used for operational planning starting conditions are necessary and must reflect exactly the current state of the facility.

Starting a facility from empty is the simplest form of starting condition. This is not uncommon; banks, shops, airports and the like all start the day without any customers in the system. Similarly, some manufacturing facilities are empty at the beginning of the day, for example an oven.

MIXED STARTING CONDITIONS AND WARM-UP

Sometimes, it is useful to have limited starting conditions which put the model into a partially representative state. The simulation is then run for a warm-up period in order to achieve a fully representative state. The hope is that the length of the warm-up period is reduced. As understanding of the model increases so the starting conditions could be modified to give a more realistic initial state, reducing the warm-up period further.

In a simulation of a storage facility I used such an approach. The contents of the store were set up in a data file and read into the model at time zero. This gave the model user control over the initial contents of the store and also reduced the length of the warm-up period, which otherwise would have included a significant time to obtain a realistic state for the store.

11.4   Multiple replications and long runs

If I were to carry out a survey in order to obtain the percentage mix of males and females in the population, I could randomly pick a sample of one hundred people. The result of the survey might show that 40 people were male and 60 female. Does this mean that 60 per cent of the population are female? How can I be certain about the accuracy of the result? By carrying out further surveys my estimate of the percentage split would change and my confidence in the accuracy of the result would improve.

In a similar way a run of a stochastic simulation model is only a sample. The results are an estimate of the real system's performance. By running the model for longer and by performing more runs the accuracy of the sample is improved. In this section methods for choosing the number of runs required and the length of the run are discussed. There is also a discussion on the preferences for multiple replications and long runs. The principles are demonstrated by the Panorama Televisions and Natland Bank case studies at the end of the chapter.

PERFORMING MULTIPLE REPLICATIONS

A single run of a simulation model is known as a replication. By changing the pseudo-random number streams and running the model again, further replications can be made. The pseudo-random number streams are changed by altering the streams referenced in the code. Most simulation software provide experimental facilities for altering the streams, enabling the user to replace one stream with another and/or to skip a specified number of values from each stream. In the latter case, care must be taken to ensure that sufficient values are skipped in order to prevent samples that were taken in a previous replication being used again. Having performed a number of replications the results are averaged in order to give a more accurate estimate of the true performance of the system being modelled (Secs 12.2 and 14.4).

How can the number of replications be determined? Two methods are discussed here; the first is a simple 'rule of thumb' and the second a graphical approach. A third statistical method is described in Sec. 14.4. It is recommended that the more detailed methods are used whenever possible.

Rule of thumb

At least three to five replications should be performed. If not, the accuracy of the results is seriously open to question and it is probably fair to say that the simulation should never have been built at all.

Graphical approach

By performing a number of replications there should be an improvement in the accuracy of the results obtained. However, each additional replication gives a smaller improvement, otherwise known as diminishing marginal returns. The question should be asked: would performing another replication significantly improve the accuracy of the results?

This question could be answered by viewing key reports such as throughput, queue sizes and customer service levels. A graph such as the one in Fig. 11.2 could be drawn. Here the cumulative mean average of a key result is plotted against the number of replications performed. If, for example, the key result were throughput and the first two replications gave a result of 100 and 150 units respectively then the first plot on the graph would be 100, the second 125 and so on. Eventually the line settles and should become nearly horizontal, at which point performing additional replications would only alter the estimate of the key result by a small amount. By visually inspecting the graph it is possible to identify the point at which this takes place and to use this as an estimate of the number of replications required; this point is shown in Fig. 11.2.

Note that if the line has settled but it is not horizontal, it is falling or rising, then more replications are required in order to obtain a horizontal line.

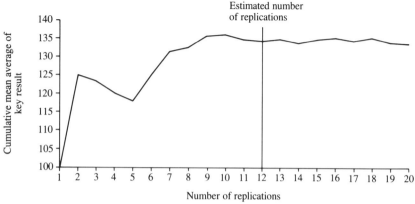

Fig. 11.2 Deciding how many replications are required: a graphical approach

Practical considerations

There are some general points to be considered when deciding how many replications to perform. The number of replications should be determined separately for each experiment. However, this may not be practical due to time constraints. Therefore, using an initial analysis to determine a fixed number of replications for each experiment may have to suffice. It is recommended that the number of replications is checked for any experiments from which conclusions are being drawn.

Obviously, for a deterministic model only one replication is possible since no pseudo-random number streams are referenced. However, if data streams (Sec. 7.5) are being used, it is recommended that the model is run with alternative input streams.

Some further points need to be considered:

- More replications need to be performed for predictive experiments since a higher level of accuracy is required than for comparative runs.
- For terminating simulations the length of run for a replication is normally determined by the termination point, while for non-terminating simulations individual judgement is required. A typical run length might be a day for a service facility and a week for a manufacturing simulation.
- In general, longer runs require fewer replications.
- If the simulation is a steady-state cycle the length of the run must be a multiple of the cycle length.

PERFORMING LONG RUNS

How long should a long run be? Two methods for deciding this are now discussed. The first is a simple rule of thumb and the second a graphical approach. It is recommended that the graphical approach is used whenever possible.

Rule of thumb

During a run, at least 10 to 20 samples should be taken from every distribution in the model. Therefore, it is necessary to consider the most infrequent events; for example, if this is the breakdown of a piece of equipment that takes place on average every five days, the length of the run, after the warm-up period, should be between 50 and 100 days.

Graphical approach

A more thorough approach is to consider the key results of the simulation. The model is warmed up and a long run performed (an initial estimate of the run length could be made using the method outlined above). Then two further replications, using different pseudo-random number streams, are made, both of the same length. If the simulation has been run for long enough the results of the three replications should be similar. This can be tested in the following way.

The cumulative mean average of the key results for each replication is drawn on a time-series such as that shown in Figure 11.3. For example, if daily throughput was being reported and for the initial replication the results of the first two days were 500 and 400 units respectively, then the first plot on the graph would be 500, the second 450 and so on. As the length of the run increases so the three lines should converge and become horizontal, indicating that the three replications are giving similar estimates for the key results; if the lines do not converge or become horizontal, the length of the run needs to be increased. The greater the level of convergence the greater the confidence in the results. Also, the longer the run the greater the convergence is likely to be; however, a point of diminishing returns is reached, beyond which convergence does not greatly improve. The run length is selected by inspecting the time-series and finding the point:

- At which the three lines have converged sufficiently to give adequate confidence in the results, for example they are within 5 per cent of each other
- Beyond which convergence, and so confidence, does not significantly improve

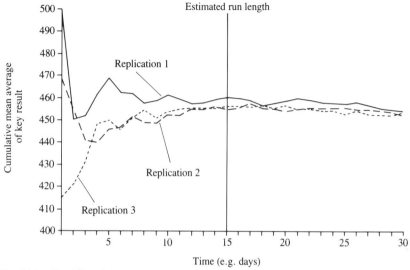

Fig. 11.3 Deciding the length of a simulation run: a graphical approach

This point is shown in Fig. 11.3. After 15 days 'adequate' confidence is attained and continuing the run would not significantly improve convergence.

Three replications, rather than two, are recommended since this significantly reduces the probability of the lines converging by chance. It is important that the lines not only converge but are also horizontal, otherwise continuing the run would cause the results to diverge, implying that the original run was too short. The interval between observations on the time-series must also be considered, and it is probably best to aim for a minimum of 10 to 20 observations over the duration of the run.

In addition to the time-series, viewing histograms of the data for each replication, for example daily throughput, can give added support to the decision regarding run length. If the histograms are not smoothly distributed then this is indicative of the run length being too short, while a smoother histogram implies a reasonable run length; the shape of the three histograms should also be similar. Example histograms are shown in Fig. 11.4.

Practical considerations

There are some general points that need to be considered when choosing the run length. The length of the run should be determined separately for each experiment; however, this may not be practical due to time constraints. Therefore, using an initial analysis to determine a fixed run length for each experiment may have to suffice, after which only a single long run is

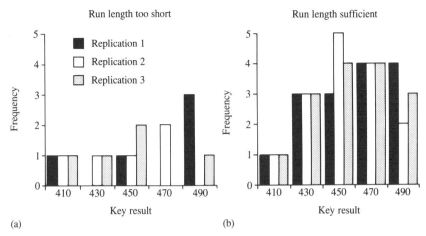

Fig. 11.4 Deciding the length of a simulation run: extension to the graphical approach

performed. It is recommended that the length of the simulation run is checked for any experiments from which conclusions are being drawn.

For deterministic models a similar analysis should be performed, though only one run is possible since no pseudo-random number streams are referenced. The run length is determined by achieving a horizontal smooth line. It is anticipated that the run length should be significantly shorter than that required for a similar stochastic simulation.

Longer runs need to be performed for predictive experiments since a higher level of accuracy is required than for comparative runs. For example, if the experiments in Fig. 11.3 were for comparative purposes only, then a run length of 10 days may suffice. If the model is a steady-state cycle simulation then the run length should be a multiple of the cycle length.

MULTIPLE REPLICATIONS OR LONG RUNS?
The question arises: is it best to perform multiple replications, long runs or both? The basic principle is the more the better, and performing long runs with a number of replications is probably best. However, it is likely that the time-scale of the project limits the number of runs that can be performed.

For terminating simulations there is no choice but to perform multiple replications, that is except for the unlikely situation in which the length of a long run is less than the termination time. For non-terminating simulations there is a choice. The advantage of performing long runs instead of multiple replications is that less time is spent warming up the model, it also gives a more realistic feel of how the facility might run over a longer period of time. The disadvantage is that statistical tests such as confidence intervals cannot be applied easily to the results of long runs since the data are not independent (Sec. 14.4).

11.5 Variance reduction

There is much discussion in the simulation literature concerning the concept of variance reduction, although in practice these concepts currently have limited application. The aim is to try and reduce the number of runs required in order to obtain an accurate result by manipulating the pseudo-random number streams, and so save experimentation time. One method of variance reduction has already been introduced in the first part of Sec. 9.2, that is the method of common random numbers.

Another popular method is the use of antithetic variates. An antithetic variate is the opposite of the values normally produced by a pseudo-random number stream. If the random numbers are between 0 and 1 then

Antithetic variates = 1 − normal variates

Hence the random numbers:

0.234, 0.976, 0.528, . . .

become

0.766, 0.024, 0.472, . . .

The idea is that if two simulation runs are performed, one with normal variates and the other with antithetic, then each should display equal but opposite effects. For example, a long breakdown which is sampled in a normal run would become a short breakdown in an antithetic run. It might seem that the average results of the two runs would give a good estimate of the true facility performance. However, in practice the idea only works for the simplest of models. The interaction of random events in more complex simulations prevents the successful use of antithetic variates.

There are a number of other methods of variance reduction such as:

- Control variates
- Selective sampling
- Descriptive sampling

These ideas have limited application in practice and require a detailed knowledge of the concepts involved. They are therefore outside the scope of this book.

11.6   Simulation and operational planning

In Sec. 1.4 (operational planning) the idea that simulation could be used for operational planning is discussed. Indeed, there are a number of applications, particularly in the area of production scheduling and maintenance planning. However, there are significant limitations to this approach at the experimental stage. It is wrong to assume that a simulation can predict the specific events of the next day. It cannot know which jobs will arrive, who will be off sick or which equipment will fail. What a simulation does is to model typical behaviour.

In operational planning a set of decisions are made; these are input to a simulation model and the model is then run. The results of this run can only be one of many possible outcomes. In order to obtain a fuller picture of the range of possible results a number of replications should be performed. However, in the operational environment time is limited and performing rigorous experimentation in this way may not be possible. Unless full experimentation is possible, simulation should probably not be used at all.

In order to overcome this difficulty deterministic models have sometimes been used in the context of operational planning. Multiple replications are no longer necessary, but the validity of the results must be open to question; a deterministic model is only one potential, and probably extreme, outcome in a complex environment in which many random events would normally occur.

However, in the absence of any other method for testing an operational plan, it could be argued that something is better than nothing, thus justifying the use of simulation.

Conclusion

In this chapter decisions concerning the warm-up, starting conditions, number of replications and run length have been addressed. There are no absolutes when tackling these issues, but the methods discussed aim to improve the decisions made. Figure 11.5 provides an overview of the experimental procedure discussed in this chapter. We now move on to discuss methods for choosing which experiments to perform and the analysis of the results.

Summary

Types of simulation model

- Deterministic and stochastic models
- Steady-state, steady-state cycle and transience
- Terminating and non-terminating simulations

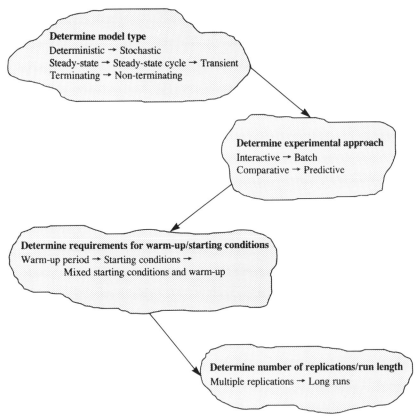

Fig. 11.5 Overview of the procedure for simulation experimentation

Methods of experimentation

- Interactive experimentation
- Batch experimentation
- Comparative experimentation
- Predictive experimentation

Warm-up and starting conditions

- Warm-up period
  - Does the state of the model represent normal working conditions (steady-state)?
  - Practical considerations: it is better for the warm-up period to be too long, save the model status at the end of warm-up and restore for each run; if time allows determine the warm-up separately for each experiment

- Starting conditions: the model starts in a representative state at time zero; the simplest form is for facilities that start from empty
- Mixed starting conditions and warm-up: partially representative state at time zero followed by a warm-up; this may reduce warm-up time

Multiple replications and long runs

- Performing multiple replications: choosing the number of replications required, rule of thumb and graphical approach
- Performing long runs: choosing the run length, rule of thumb and graphical approach
- multiple replications or long runs?: the more the better; depends on the project time-scale; for terminating simulations multiple replications are the only option; long runs are recommended for non-terminating simulations

Variance reduction

- Methods: common random numbers, antithetic variates
- Limited application

Simulation in operational planning
- Limited application

Case study 11.1: Panorama Televisions

Model type
The Panorama Televisions model has the following characteristics:

- Stochastic
- Non-terminating
- Steady-state

Since the production schedule is continuously repeated it could be argued that the model is steady-state cycle. However, the schedule only affects the frequency of set-ups and there is no intention to experiment with alternatives; also, the schedule is repeated over a short period. Therefore, the model may to all intents and purposes be treated as a steady-state simulation.

The experiments are for predictive purposes (see the example's objectives in Chapter 4) and so a high degree of accuracy is required.

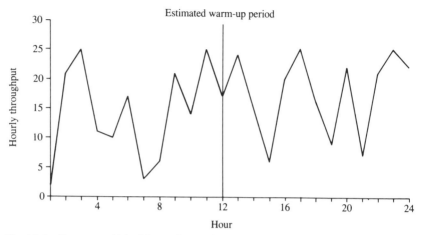

Fig. 11.6 Panorama Televisions: time-series of hourly throughput to determine the warm-up period

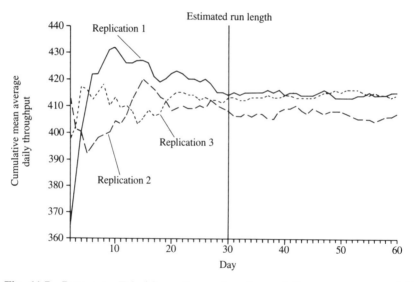

Fig. 11.7 Panorama Televisions: time-series of cumulative mean average daily throughput to determine the run length

Warm-up period

In order to determine the warm-up period required the key results of work-in-progress and throughput have been considered. Since this is a palletized system the work-in-progress builds to a maximum of 30 pallets and then never changes. By watching the model run it was noted that the

Table 11.1 Panorama Televisions: daily throughput and cumulative mean average daily throughput to determine the run length

Day	Replication 1		Replication 2		Replication 3	
	Daily throughput	Cumulative mean average daily throughput	Daily throughput	Cumulative mean average daily throughput	Daily throughput	Cumulative mean average daily throughput
1	371	371	426	426	405	405
2	359	365	401	414	388	397
3	414	381	375	401	412	402
4	461	401	397	400	462	417
5	453	412	363	392	415	416
6	476	422	409	395	387	412
7	417	422	414	398	437	415
8	458	426	405	399	435	418
9	470	431	406	400	347	410
10	441	432	441	404	437	413
.	.	.	.	.	.	.
.	.	.	.	.	.	.
.	.	.	.	.	.	.

work-in-progress reached its maximum in approximately one hour. Figure 11.6 shows the hourly throughput over the first day of a model run. The throughput soon reaches a representative level.

A warm-up period of 12 hours has been selected. This is probably longer than required (one hour may well suffice) but it does ensure that any initial bias is removed. In addition, this is a relatively small overhead in comparison to the run length.

Run length
Three replications, with different pseudo-random number streams, have been performed with a 12 hour warm-up period followed by a 60 day run. The results of the first 10 days are summarized in Table 11.1 and the cumulative mean average daily throughput for the full set of results is shown in Fig. 11.7.

Note that the cumulative mean average daily throughput is calculated as follows:

$$\frac{\text{Throughput day}_1 + \cdots + \text{throughput day}_n}{n}$$

where n is the number of days over which the average is being calculated.

For example, the cumulative mean average daily throughput at the end of day 3 for replication 1 is

$$\frac{371 + 359 + 414}{3} = 381$$

By visually inspecting Fig. 11.7 a run length of 30 days has been selected. At this point the results are within 2 per cent of each other and continuing the run would appear to give little improvement to the confidence in the results obtained. (Note that if the model was being used for comparative purposes only, a run length of, say, 20 days may well suffice.) This run length is further confirmed by Fig. 11.8 in which the distributions of daily throughput for the first 30 days of each replication are shown. All three distributions are relatively smooth and are reasonably similar.

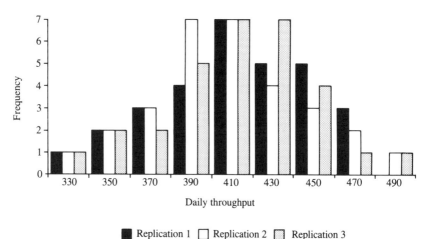

Fig. 11.8 Panorama Televisions: distributions of daily throughput for three replications

Conclusions
Experiments with the Panorama Televisions simulation will be run with:

- A 12 hour warm-up period
- One run of 30 days' duration

Case study 11.2: Natland Bank

Model type
The Natland Bank model has the following characteristics:

- Stochastic
- Terminating
- Steady-state cycle (cycle length of one day based on customer arrival rate)

The experiments are largely for comparative purposes (see the example's objectives in Chapter 4) and so the degree of accuracy need only suit this purpose.

Warm-up period/starting conditions
The bank opens at 9.30 a.m. with no customers queuing. The model represents this starting condition exactly from the beginning of a run. Therefore, no warm-up period is required.

Run length
Since this is a terminating simulation the run length is already determined at one day. In practice the model will be run for the six hours of opening time and then the run continued until all the remaining customers have left the bank. The accuracy of the result will be increased by performing a number of replications with different pseudo-random number streams.

Number of replications, graphical approach
Twenty replications have been performed with two tellers and two queues. The mean average queuing time is the key result, and so the cumulative mean average queuing time has been calculated in Table 11.2 and shown in Fig. 11.9. It is estimated that six replications are required since the graph is reasonably settled at this point. (If the model were being used for predictive purposes more replications would probably be necessary, say 12, in order to increase the confidence in the result.)

Note that the cumulative mean average queuing time is calculated as follows:

$$\frac{\text{Average queuing time replication}_1 + \cdots + \text{Average queuing time replication}_n}{n}$$

where n is the number of replications over which the average is being calculated. For example, the cumulative mean average queuing time after three replications is

Table 11.2   Natland Bank: calculation of cumulative mean average queuing time

Replication	Mean average queuing time (min)	Cumulative mean average queuing time (min)
1	2.27	2.27
2	1.57	1.92
3	1.49	1.78
4	2.86	2.05
5	2.29	2.10
6	3.30	2.30
7	2.18	2.28
8	1.64	2.20
9	1.02	2.07
10	3.92	2.25
11	1.79	2.21
12	4.07	2.37
13	1.92	2.33
14	1.01	2.24
15	1.68	2.20
16	1.92	2.18
17	2.09	2.18
18	1.22	2.12
19	1.24	2.08
20	3.98	2.17

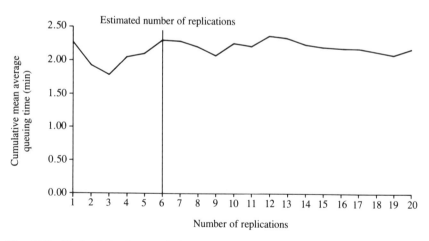

Fig. 11.9   Natland Bank: cumulative mean average queuing time to determine the number of replications

$$\frac{2.27 + 1.57 + 1.49}{3} = 1.78$$

Conclusions
Experiments with the Natland Bank simulation will be run with:

- No warm-up period
- A run length of 1 day
- Six replications for each experiment

12

Performing experiments and analysing the results

Performing experiments can be a very time consuming and sometimes stressful business. Regularly I have set up experiments and left computers running before going home in the evening. At weekends I have visited the office to ensure that the experiments are still running and that there has not been an error in the model or a power cut. I have even taken computers home and run experiments in order to keep a close watch on what is happening. On a number of occasions I have had a client eagerly awaiting the results first thing on a Monday morning. It is somewhat difficult to explain when the experiments have failed for whatever reason. In some cases experiments have been a wasted effort and could not achieve the purpose for which they were run; with more thought they need never have been performed. Other experiments have come close to achieving some target, but have also proved fruitless.

Providing a proper focus and aiming to contain the number of experiments that need to be performed is vital if the project is to be kept on time. Part of this process is analysing the results, drawing sensible conclusions and making recommendations. It is the aim of this chapter to show how these issues can be addressed. Firstly, the selection of experiments is discussed followed by an overview of results analysis. The chapter concludes by performing experiments with the Panorama Televisions and Natland Bank case studies and analysing the results.

12.1 Performing experiments
In the previous chapter methods for deciding the warm-up period, run length and number of replications are discussed; it is now time to consider which experiments should be performed. Despite the fact that these decisions have been split across two chapters they are by no means separate

phases of work. In fact, the decisions regarding warm-up, run length and replications should continually be reviewed throughout this phase of the project.

In order to aid the process of experimentation most simulation packages provide facilities for setting up experiments and leaving them to run overnight. As a minimum this should include the ability to load models, change pseudo-random number streams, batch forward in time and save the results as well as the model status. Experiments are either set up using batch files to control the model or the software may have its own experimental features.

There are broadly two approaches that can be taken to the selection of experiments:

1 *Comparing alternatives* When comparing alternatives there are a limited set of options that need to be compared against one another using some decision criteria. The number of alternatives is normally small; however, there are occasions when a large number exist. Typical examples of this type of experimentation are: confirming that a facility achieves target throughput (one alternative only), comparing two alternative facility configurations and evaluating customer service under five alternative control policies. It is likely that the set of alternatives were either known when the objectives of the project were defined or they have evolved as the study has progressed. This is a common approach to experimentation in simulation projects.

2 *Searching for a target/optimum* In this form of experimentation there are no predetermined alternatives to be compared. However, there are one or more experimental factors that need to be varied until a target or optimum, based on some decision criteria, is reached. Examples are: changing the size of buffers until target throughput is reached and altering staff rosters in order to achieve maximum customer satisfaction. There is potential for a large number of experiments since the experimental factors can be set at many levels and there may be a number of factors that impact on the performance of the facility. The number of experiments can be calculated as follows:

$$\text{Number of experiments} = \text{number of levels}^{\text{number of experimental factors}}$$

For example, if there were three experimental factors and each could be set at two levels there would be 2^3, that is 8, experiments. If there were four experimental factors and three levels there would be potential for 3^4, that is 81, experiments. This means that the choice of experiments needs to be considered carefully in order to avoid the completion date of the project disappearing into the distance.

The rest of this section now focuses on how to approach search experimentation, selecting the correct experiments and reducing the number required. Some of these ideas may also be useful for comparing alternatives when there are a large number of options to be considered and full experimentation is not possible within the time-scale of the project.

DETERMINE THE IMPORTANT EXPERIMENTAL FACTORS

When starting search experimentation it is often useful to determine the relative importance of each experimental factor, especially when there are a large number. Which factors are important, giving a large improvement towards the objective, and which are less important? For instance, increasing the size of buffers may have a significant impact on throughput while changing cycle times has only a small impact. It is also important to estimate the expected size of this impact. Having done this, experimentation can focus on the important factors and so reduce the total number of experiments required. There are three ways of analysing the importance of the experimental factors:

1 *Data analysis* By analysing the data in the model it is often possible to draw conclusions about the impact of the experimental factors on the performance of the facility. Such analysis is obviously limited in that it cannot take account of complex random events and their interaction, but it can give an indication. For example, by comparing the average cycle time and percentage up-time across various pieces of equipment the operations that need most attention can probably be identified.
2 *Expert knowledge* Facility experts are often able to identify the relative importance of various experimental factors and it is useful to interview them in order to obtain this information. However, it is recommended that further analysis is performed in order to ensure that their conclusions are correct and to assess the actual size of the impact.
3 *Preliminary experimentation* Preliminary experiments can be performed for the purpose of measuring the effect of each experimental factor. This analysis could be carried out in a similar way to the sensitivity analysis described in Sec. 10.4.

The advantage of data analysis and expert knowledge is that they require less time than preliminary experimentation. However, preliminary experimentation normally provides a more thorough investigation.

In all of this the practicality and cost of making the change should be considered. Although an experimental factor may have a significant impact on the performance, making the change may be less practicable and more expensive than an alternative approach. Such an experimental factor would

Table 12.1   Factorial design of experiments

Experiment number	Level of experimental factor 1	Level of experimental factor 2	Level of experimental factor 3
1	1	1	1
2	1	1	2
3	1	2	1
4	1	2	2
5	1	3	1
6	1	3	2
7	2	1	1
8	2	1	2
9	2	2	1
10	2	2	2
11	2	3	1
12	2	3	2

be considered less important than another factor that had the same impact but at a lower cost. For example, if improving maintenance procedures has a similar impact to increasing buffer sizes but proves less costly and more practicable, then this factor should be seen as more important.

EXPERIMENT WITH THE IMPORTANT EXPERIMENTAL FACTORS
Having decided which experimental factors are important and the expected size of their effect, it should be possible to identify a number of initial experiments that are likely to yield useful results. On some occasions these initial experiments achieve the target or optimum performance and no further experimentation is required. On other occasions they provide useful information which can then be used to identify further experiments; an iterative process of discovery begins, performing experiments, analysing the results and identifying further experiments in order to search for the target or optimum.

Factorial design

A set of experiments is made up of a number of experimental factors which are set at various different levels. This is known as a factorial design and an example is shown in Table 12.1.

Fractional factorial design

Where a large number of experiments need to be performed, project time-scales may prevent a full evaluation of all the alternatives. In this case a

Experimental factor 2

		Level 1	Level 2	Level 3
	Level 1	A	b	C
Experimental factor 1	Level 2	d	E	f
	Level 3	G	h	I

Fig. 12.1 Simplified fractional factorial design of experiments

fractional factorial design could be used; a limited number of experiments are performed and conclusions are drawn regarding the likely outcome of the other experiments. Statistical methods have been developed for performing such an analysis but these are outside the scope of this book; they are well reported in the simulation literature. However, a practical approach is now discussed.

The number of experiments can be reduced by testing a limited number of alternatives and using visual inspection and judgement to draw conclusions regarding experiments that have not been performed. Take an example where there are two factors each of which are to be set at three different levels. The experiments are shown in Fig. 12.1. Rather than performing all nine experiments, runs could be performed for experiments A, C, E, G and I, giving the extreme values and the mid-point E. The key results could then be shown in the table. By visually inspecting this table it should be possible to draw conclusions about the likely outcomes of the experiments b, d, f and h. If it is suspected that one of these would meet the project's objectives then the experiment should be performed in order to confirm, or deny, this suspicion.

INFINITE QUEUE METHODS

If one of the experimental factors is the maximum size of the queues or buffers the experimentation can be simplified further. Instead of setting the queue sizes to certain levels and performing various experiments, it is quicker to allow the queues to float to any size—infinite queues. One experiment is run and the maximum queue sizes reported. This maximum is then taken as an indication of the queue requirements.

Even if there are physical limitations to the size of queues such experimentation may still prove useful in identifying the limits of the facility's performance.

12.2   Results analysis

Throughout the process of experimentation, as well as on its completion, it is important to analyse the results that are being obtained. The aim is to check whether the objectives of the project have been met and the extent to which they have been achieved. This enables conclusions to be drawn from the experiments which have been carried out and also helps to identify any further experiments that need to be performed. Having completed the experimentation and results analysis a set of conclusions and recommendations should become apparent. In this section the need to analyse the results is discussed.

GENERAL PRINCIPLES OF RESULTS ANALYSIS

The results can be analysed with reference to two types of report: point estimates and measures of spread. Since simulation experiments produce only a sample of the potential results it is important to remember that these are only estimates of the true facility's performance. Some simulation packages provide facilities for analysing results while the use of third party software, such as spreadsheets or simulation analysis packages, can further enhance this process.

Point estimates

These are an estimate of the average or aggregate performance of the facility, for example:

- Cumulative total and percentage
- Mean
- Median
- Mode

When multiple replications have been performed an average for all the replications should be provided as well as the spread of the point estimates, for instance a histogram or the standard deviation. In this way a greater understanding of the true value of the facility performance is obtained. For example, if the results of three replications give a throughput of 540, 580 and 530 units, the mean average and the standard deviation could be calculated, 550 and 26.5 units respectively. On the other hand, from another model results of 420, 690 and 540 may be obtained, giving a mean of 550 units but a standard deviation of 135.3. In the first model the point estimates from the replications have a smaller spread; therefore, there is more confidence concerning the true value of the performance. This is the first step in constructing a confidence interval (Sec. 14.4).

Measures of spread

These are an estimate of the variation in the performance of the facility, for example:

- Standard deviation
- Quartiles
- Minimum and maximum
- Histograms

These results are distinct from the spread of the point estimates discussed above since they show the expected variation about the point estimate. An example is a histogram showing the queuing time for every individual customer in a service system.

COMPARING RESULTS

Before performing any experiments it is useful to decide which criteria should be used to judge the performance of the facility. What is good performance and what is poor performance? This should have already been discussed while defining the objectives of the project (Chapter 4) and the model's reports (Chapter 5).

For every criteria the point estimates and, where applicable, the measures of spread should be compared.

Comparing point estimates

Since the results are only based on samples of the true facility performance care must be taken when comparing point estimates. It does not necessarily follow that an experiment that yields a throughput of 1026 units is better than one that yields 1025 units. The question must be asked: what constitutes a significant difference? Three factors must be considered in answering this. Firstly, there is the actual size of the difference. A 10 per cent improvement is highly likely to be significant, while a 0.001 per cent improvement is almost certainly insignificant. Secondly, t ere is the size of the sample taken. The simulation result becomes more accurate with more replications and longer experimental runs. Therefore, a small improvement has greater significance when more thorough experimentation has been performed. Thirdly, there is the relative spread of the point estimates obtained from the replications. A wider spread implies that there is less certainty regarding the accuracy of the point estimates in predicting the true value of the result. Again, by performing more replications the accuracy of the result is improved. By considering these three factors and using individual judgement it should be possible to decide whether one result is significantly different from another or whether more thorough experimentation is required before a judgement can be made. Alternatively, statistical methods can be used; these are discussed in Sec. 14.5.

Comparing measures of spread

The variability in the results, as well as the point estimates, should be compared. Take the extreme example of the histograms showing daily throughput in Fig. 12.2. Although the mean performance of the facilities is the same, the variation is somewhat different. In most situations the first histogram, with lower variation, would be the preferred option. Variability is an important factor in comparing results. It may be that although an experiment gives an improvement in the point estimate the result is considered worse due to poorer variability. Predictable performance is often preferred to a higher average with greater variability.

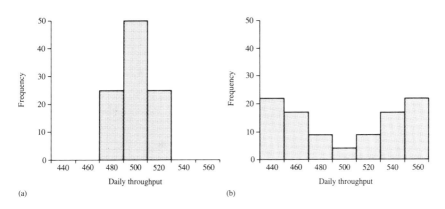

Fig. 12.2   Comparing measures of spread

DRAWING CONCLUSIONS AND MAKING RECOMMENDATIONS
The conclusions which are drawn from the results of an experiment are open to all kinds of misinterpretation. In one simulation project I was able to demonstrate the significant difficulties that a production facility would have in meeting the demand for sub-components from another plant. There were two potential solutions: one was to increase the level of safety stocks and the other was to schedule production correctly. I think it is obvious which option incurs least expense. Having presented the

results to the plant director the message seemed to become somewhat confused. A few days later, to my horror, I discovered that the director had issued a request for safety stocks to be doubled! Suffice it to say I had to correct this mistake.

The principle is that the results and their implications should be considered carefully before any conclusions are drawn and recommendations made. It is not possible here to give exact guidance on how this should be carried out since the methods used depend upon the project, the skills of the project team and the needs of the organization. However, it is important to remember that the simulation is primarily a decision support tool and that the results should be considered in the light of the organization as a whole. The costs and benefits of each option, both quantitative and qualitative, need to be discussed and the project team should give thorough consideration to every alternative before any recommendations are made.

12.3 Sensitivity analysis

In Sec. 10.4 there is a discussion on the need to perform a sensitivity analysis on any unobtainable (category C) data during model validation. If the model was found to be sensitive to these data it is recommended that further analyses should be performed during experimentation and a range of expected results reported. Now is the time to perform these analyses before the final conclusions are drawn and any recommendations are made.

The sensitivity analysis could also be extended to other data in order to test the limits within which the solution applies. No data are absolutely certain and they could change between the completion of the simulation project and implementation. A good solution is also robust. If a small change to the data alters the solution then this should be considered when the recommendations are being made.

The potential for sensitivity analysis is almost limitless and time is not. It is therefore recommended that such an analysis is restricted to the unobtainable data and a few other key factors.

Conclusion

The simulation experiments have been run and the results analysed to draw conclusions. All that remains is to present the results and implement them. These are part of the project completion and implementation phase.

Summary

Performing experiments

- Comparing alternatives

- Searching for a target/optimum
- Determine the important experimental factors: data analysis, expert knowledge, preliminary experimentation
- Experiment with the important experimental factors: factorial design, fractional factorial design
- Some simulation packages provide experimental frameworks/tools
- Infinite queue method of experimentation

Results analysis

- Point estimates: cumulative total and percentage, mean, median, mode
- Measures of spread: standard deviation, quartiles, minimum and maximum, histograms
- Provide averages of point estimates and a measure of their spread for multiple replications
- Comparing results: point estimates are only estimates; need to establish whether any differences are significant; consider variability as well as point estimates
- drawing conclusions and making recommendations: discuss costs and benefits, quantitative and qualitative

Sensitivity analysis

- Perform further sensitivity analysis on unobtainable (category C) data
- Test the limits of other data within which the solution still applies

---

Case study 12.1: Panorama Televisions

The experimentation is aimed at achieving a target throughput of 500 units per day. For all the experiments outlined below one run of 30 days has been performed with a 12 hour warm-up period (Chapter 11).

Experimentation with additional pallets
Three experiments have been performed to test the effect of increasing the number of pallets. The results are shown in Table 12.2. Increasing the number of pallets by 50 per cent has some effect; however, increasing the number beyond this does not significantly increase the daily throughput. Additional buffering must be considered.

Initial experimentation with additional pallets and buffering
Since the effect of additional buffering is not known, two initial experi-

Table 12.2 Panorama Televisions. Experimental results: additional pallets

Number of pallets	Mean daily throughput
30 (current level)	414.17
45	430.47
60	431.00

Table 12.3 Panorama Televisions. Experimental results: additional pallets and buffering

Percentage increase in buffers	Number of pallets	Mean daily throughput
0	45	430.47
100	100	439.93
200	150	440.60

Table 12.4 Panorama Televisions. Experimental results: machine utilizations from 200 per cent increase in buffers

Operation	Percentage idle	Percentage busy	Percentage blocked	Percentage set-up	Percentage broken	Percentage waiting for labour
OP10	0.00	58.06	41.94	0.00	0.00	0.00
OP20	0.00	64.23	20.79	5.55	5.58	3.85
OP30	1.23	61.19	22.63	5.60	6.04	3.30
OP40	1.76	61.20	37.04	0.00	0.00	0.00
Test	0.09	48.29	24.67	9.75	6.80	10.40
Re-work	42.24	57.76	0.00	0.00	0.00	0.00
OP50	2.80	64.25	0.00	15.73	5.96	11.25
OP60	41.85	58.15	0.00	0.00	0.00	0.00

ments have been performed to investigate the impact. In the first, the buffering was increased by 100 per cent and in the second by 200 per cent. The number of pallets has been increased to 100 and 150 respectively. The results are shown in Table 12.3.

Increasing the buffers by 100 per cent improves daily throughput by 10 units; however, a further increase seems to have no significant effect. This is a surprising result and points to some bottleneck in the facility. Table 12.4 shows the results obtained for machine utilization in the experiment with a 200 per cent increase to the buffers.

Investigation of Table 12.4 shows that OP50, back assembly, is the bottleneck operation with very little idle or blocked time. A major cause of lost production is OP50 waiting for repair/set-up labour, as indeed is the case for some of the other operations. Therefore, an experiment was performed with two, rather than one, repair/set-up labour. This demonstrates the introduction of a new experimental factor previously not considered. The results, shown in Table 12.5, demonstrate a significant improvement in throughput by employing an additional repair/set-up operator.

Full experimentation with additional pallets and buffering
Having established the need to employ an additional repair/set-up operator full experimentation with additional pallets and buffering was performed. The aim of these experiments was to obtain a detailed understanding of the relationship between the daily throughput, the number of pallets and the buffering. The purpose was also to see whether the target throughput could be achieved within the limits of the experimental factors, namely a 200 per cent increase in buffering. The factorial design and the results are shown in Table 12.6. For all these experiments two repair/set-up operators were employed.

Conclusions and recommendations
The following conclusions have been drawn from the above results:

- An additional repair/set-up operator is required
- Using the correct number of pallets and increasing the buffering has a significant effect on throughput
- A throughput of 490 units per day can be achieved with a 200 per cent increase in buffers and 125 pallets
- It is not possible to achieve target throughput by increasing the number of pallets and quantity of buffering (within defined limits) alone

Since the target throughput has not been achieved additional policies should be considered. For example, increasing the buffering by more than 200 per cent, improving maintenance to reduce the number of breakdowns and changing the production schedule to reduce the number of set-ups. It is probable that the latter two policies would reduce the need for additional buffering and pallets. However, there is some doubt concerning Panorama's ability to control these factors (Chapter 4). This work is outside the scope of this project and should be considered as part of a separate study.

Table 12.5 Panorama Televisions. Experimental results: additional repair/set-up labour

Percentage increase in buffers	Number of pallets	Number of repair/ set-up labour	Mean daily throughput
200%	150	1	440.60
200%	150	2	489.33

Table 12.6 Panorama Televisions. Factorial design and results of full experimentation

Experiment number	Number of pallets	Percentage increase in buffers	Mean daily throughput	Standard deviation
1	50	0	460.10	35.51
2	75	0	447.40	35.55
3	50	100	461.60	33.23
4	75	100	477.97	38.16
5	100	100	478.23	38.23
6	125	100	478.03	35.70
7	150	100	407.10	40.09
8	100	200	487.93	30.88
9	125	200	489.33	31.83
10	150	200	489.33	31.83
11	175	200	488.50	30.58
12	200	200	478.50	36.03

Case study 12.2: Natland Bank

Experiments performed
The objectives of the Natland Bank project describe a set of alternatives that need to be compared, namely:

- 1 teller, 1 queue
- 2 tellers, 1 queue
- 2 tellers, 2 queues
- 3 tellers, 1 queue
- 3 tellers, 3 queues

For each of these experiments six replications have been performed (Chapter 11).

Table 12.7   Natland Bank. Experimental results: one teller, one queue

Replication	Percentage of customers within 3 min
1	6.51
2	4.44
3	3.38
4	3.69
5	9.20
6	2.00
Mean	4.87
Standard deviation	2.59

Table 12.8   Natland Bank. Experimental results: two tellers, one queue

Replication	Mean queuing time (min)	Percentage of customers within 3 min	Teller utilization (%)
1	1.87	76.69	77.39
2	1.74	78.27	80.73
3	1.43	80.75	79.13
4	2.76	64.80	83.75
5	2.58	66.82	81.74
6	3.16	56.94	81.79
Mean	2.26	70.71	80.75
Standard deviation	0.67	9.31	2.24

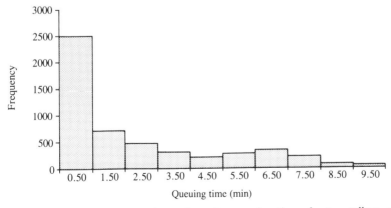

Fig. 12.3   Natland Bank. Experimental results: queuing times for two tellers, one queue

Table 12.9 Natland Bank. Experimental results: two tellers, two queues

Replication	Mean queuing time (min)	Percentage of customers within 3 min	Teller utilization (%)
1	2.27	67.41	76.72
2	1.57	81.14	80.46
3	1.49	80.21	79.15
4	2.86	62.67	83.25
5	2.29	74.33	81.09
6	3.30	56.18	81.43
Mean	2.30	70.32	80.35
Standard deviation	0.71	9.98	2.23

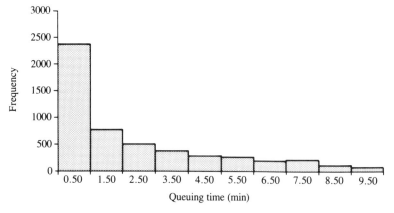

Fig. 12.4 Natland Bank. Experimental results: queuing times for two tellers, two queues

Table 12.10 Natland Bank. Experimental results: three tellers, one queue

Replication	Mean queuing time (min)	Percentage of customers within 3 min	Teller utilization (%)
1	0.11	100.00	52.35
2	0.10	100.00	53.41
3	0.10	100.00	52.84
4	0.13	100.00	56.73
5	0.14	100.00	54.27
6	0.15	100.00	56.13
Mean	0.12	100.00	54.29
Standard deviation	0.02	0.00	1.79

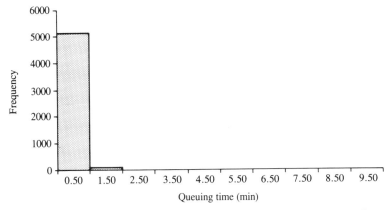

Fig. 12.5 Natland Bank. Experimental results: queuing times for three tellers, one queue

Table 12.11   Natland Bank. Experimental results: three tellers, three queues

Replication	Mean queuing time (min)	Percentage of customers within 3 min	Teller utilization (%)
1	0.32	100.00	51.37
2	0.33	100.00	53.51
3	0.32	100.00	52.82
4	0.36	100.00	56.37
5	0.36	100.00	53.95
6	0.39	100.00	55.95
Mean	0.35	100.00	53.99
Standard deviation	0.03	0.00	1.89

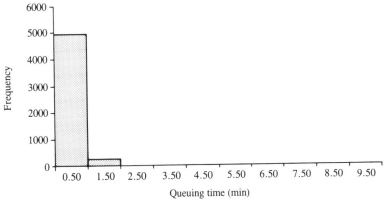

Fig. 12.6 Natland Bank. Experimental results: queuing times for three tellers, three queues

Experimental results and analysis

A summary of the key results for each experiment are given in Tables 12.7 to 12.11 and histograms of the queuing times for all replications are shown in Figs 12.3 to 12.6. For the one teller, one queue experiment the results showing the percentage of customers within three minutes give such poor performance that no further analysis has been carried out.

Conclusions and recommendations

The following conclusions have been drawn from the above results:

- One teller is not capable of meeting customer service requirements
- With two tellers 70 per cent of customers are served within three minutes of arriving at a queue, although some customers have to wait for up to 10 minutes at peak times
- With three tellers all customers are served within three minutes of arriving at the auto teller queues
- Teller utilization is on average 80 per cent with two tellers and 55 per cent with three
- A single queue gives a slight improvement in service with three tellers but for two tellers there is no significant difference; this is due to the level of queue jockeying

Two or three auto tellers should be purchased. The improvement in customer service with three auto tellers needs to be weighed against the additional cost. A single queue may give some slight improvement in customer service and should therefore be considered.

PART FIVE

Project completion and implementation

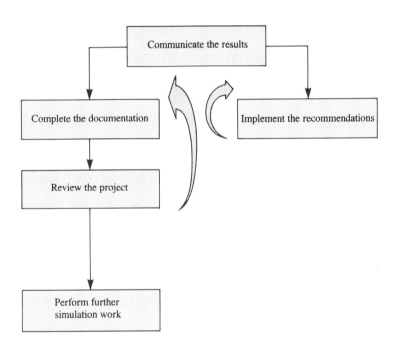

13

Project completion and implementation

Many celebrated artists have lived the duration of their lives in relative obscurity and often in poverty, their talent not recognized until some time after their death. Vincent Van Gogh sold only one painting during his life, and that was to his brother! If only he could enter a modern day auction room and see his works selling for inordinate sums of money, even millions. There is some perverse sense of justice in the fact that he is finally being recognized although it is of little use to Van Gogh himself. As for why such artists never received acclaim in their own lifetime, there could be many reasons. They may not have attempted to market themselves, or possibly did not have the skills to do so; perhaps the message they portrayed was not a popular one; they probably did not meet with contemporary tastes; or perhaps they were simply never in the right place at the right time. The reasons are of little importance here but the fact is that excellent work is of little value until it is recognized by others, except, that is, for providing personal satisfaction. Without recognition it is almost impossible for a piece of work to be considered a success.

Thankfully, simulation modellers do not have to die before their work can receive acclaim, or at least we hope that this is not the case! However, there is a real danger that the results of a simulation study may fall into relative obscurity through lack of attention to the final phase of the project, completion and implementation. The project cannot be considered a success until this has been carried out, and carried out well.

It is the intention of this chapter to give some guidance on how to complete a project and to implement the findings. Firstly, the results, including the conclusions and recommendations, must be communicated; secondly; they need to be implemented. While this is being done the project documentation should be completed, the project reviewed and any further simulation work progressed. Each of these are discussed in turn and the chapter concludes by giving summary reports for the Panorama Televisions and Natland Bank case studies.

13.1 Communicating the results

The results, including the conclusions and recommendations, of the project need to be communicated to each member of the project team and to all other interested parties. This includes any relevant managers and those who are represented by members of the project team.

WHY COMMUNICATE THE RESULTS?

There are a number of reasons for communicating the results:

1 *Client satisfaction* The client has committed resources in both time and money to the project and would therefore like to see some payback. Although the project may have given significant benefits, unless these are communicated the perception of the client may be somewhat different. The project and simulation may be seen as a cost with little benefit and there is a danger that both will fall into disrepute. The results of the project need to be communicated in order to ensure that the client is satisfied. This also increases the chance that other simulation projects will be agreed, whether this is further work on the same project or a completely new study. If it is a large project then opportunity should be taken to communicate the results at regular intervals in order to maintain client interest.

2 *Developing new ideas* When the results are communicated to the members of the project team and the management, they may be able to suggest new approaches for tackling the issues that are raised. Communication becomes a catalyst for new ideas and part of an iterative process of experimentation and reporting back.

3 *Obtaining agreement to the recommendations* The recommendations of the project need to be communicated, discussed and agreed in order to gain support before moving to implementation. This is the most important reason for communicating the results of the project.

METHODS OF COMMUNICATING THE RESULTS

The results may be communicated by written report and verbal presentation. A presentation can be given as a formal session or as an informal discussion. The method chosen largely depends upon the needs of the project and the organization. The advantages of each are similar to those discussed for the project specification (Sec. 8.3).

The aim of the report or presentation is to provide a summary of the project. In the report, detailed figures and tables showing the results of experiments should be provided in an appendix and only the key results given in the main body. As a guide, the content of the report or presentation should be as follows (the chapters or sections where the information is discussed are shown in brackets):

- Executive summary: objectives, key results, conclusions and recommendations
- Introduction to the problem and project objectives (Chapter 4)
- Model summary (Chapters 5 and 6)
- Outline of experimentation (Chapters 11 and 12)
- Key results (Sec. 12.2)
- Conclusions and recommendations (the last part of Sec. 12.2)
- Further work (Sec. 13.5)
- Appendices: detailed results

It may also be useful to provide a summary of the benefits obtained and the objectives achieved. There is some repetition of the project specification (Chapter 8) since this may not have been read by some, and it is useful to remind others of the project's objectives and the details of the model. If the results are being presented it may prove useful to demonstrate some of the key findings with the model. However, care must be taken that the model demonstration does not detract from the purpose of the session, to communicate the results and obtain agreement for the recommendations.

Having communicated the results there should be opportunity for feedback. In this way any doubts can be aired and misunderstandings explained. Above all, the purpose is to ensure that the results and conclusions are understood and that the recommendations are agreed before moving to implementation.

13.2   Implementing the recommendations

On a number of occasions I had explained that there was a serious flaw in the design of a manufacturing facility. Further to this, the simulation had shown how a simple change to the design could dramatically improve throughput. Initially this modification was incorporated in the layout. However, a few months later I discovered that the change had been neglected and was no longer included in the design. Having explained the reasoning afresh the modification was introduced again. In fact, this process of designing-in and designing-out the modification, between explanations of why it was there in the first place, continued for some time. The implementation of this recommendation was somewhat deficient.

How can the recommendations be implemented successfully? It is important to:

- Identify who will implement the recommendations
- Develop an implementation plan
- Monitor the outcome

Implementation should be seen as a project in its own right. Each of the above are now discussed.

IDENTIFY WHO WILL IMPLEMENT THE RECOMMENDATIONS

Firstly, it is important to identify who will carry out the implementation. Is it the simulation project team, a separate team formed by the client or a joint effort? The advantage of keeping at least some members of the project team involved is that they have an in-depth understanding of the recommendations. If any difficulties arise during the implementation, and further simulation work is required, then modelling expertise is close at hand. However, the disadvantage is that their skills may be required for another simulation project which will consequently be delayed until the implementation is complete. If the client forms a separate team then the simulation expertise is released for other work, but it is important to ensure that the new team has a full understanding of the recommendations. A joint team may prove a useful compromise.

A project team should be set up in order to see the implementation through to completion. In a similar way to the simulation project team (Sec. 2.5), this group may need to consist of many members or of only one, depending on the effort and skills required to carry out the implementation.

Having identified the members of the implementation project team there begins a process of education. Each member needs to fully understand the recommendations in order to ensure that they can implement them properly. The involvement of the simulation project team is at least required for this stage.

DEVELOP AN IMPLEMENTATION PLAN

Now that the members of the implementation project team have been identified and they understand what is required, an implementation plan should be developed. This should identify:

- The recommendations to be implemented
- How the recommendations will be implemented, including project phases
- Individual responsibilities
- When the implementation (of each phase) will be complete

The emphasis is practical, that is on how the recommendations can be made to work. Since the simulation is necessarily an approximation to reality, some adjustment to the recommendations may be required in order to ensure that they work in practice.

It is probably best to have a project specification, in the same way that one was provided for the simulation project (Chapter 8). This should outline the information discussed above and acts as a formal record of the implementation project. In this way, greater control over implementation is achieved.

MONITOR THE OUTCOME

Once the plan has been put into practice the outcome should be monitored by comparing the performance of the real facility against the results of the simulation. Hopefully the results compare favourably.

However, differences do occur. This could be the result of a faulty implementation, in which case the reasons for this need to be reviewed and corrective action taken. Alternatively, the conditions under which the facility is operating may have changed and the original recommendations are less, or no longer, valid. In order to avoid this difficulty during implementation, whenever conditions change, the simulation could be updated and the validity of the recommendations checked.

The third reason why differences occur is that the simulation is not valid, although this is unlikely when rigorous model validation has been performed (Chapter 10). If this proves to be the case, the reasons need to be uncovered in order to prevent the same from happening in future.

HANDING OVER A MODEL

Sometimes the implementation consists simply of delivering a simulation tool. This occurs when the simulation is to be given to a model user for experimentation (Sec. 2.5). In these circumstances user documentation (Sec. 9.2) should be provided and possibly a training session. Following hand-over, continued support should be made available to the model user via a telephone help-line and/or allocated man-time.

13.3 Completing the documentation

Throughout the project various documents have been created, namely (the chapters or sections where the information is discussed are shown in brackets):

- Project specification (Chapter 8)
- Model documentation (Sec. 9.2)
- User documentation (Sec. 9.2)
- Final report (Sec. 13.1)
- Minutes of project team meetings (Sec. 2.5)
- Implementation project specification (Sec. 13.2)
- Project review (Sec. 13.4)

It is important to ensure that all of these are complete and are readily available for further use. There may be a requirement for the simulation to be used again at some future date and thorough documentation enables the model to be understood more easily, especially if a new modeller or model user is involved. This documentation also provides a useful history of the project and acts as supporting information when the project is being reviewed.

13.4 Reviewing the project

The time-scale of one project had slipped by a significant amount, which led to a number of sleepless nights and quite justifiable complaints from the client. Thankfully, disaster was averted, useful results were obtained and the client pacified. Following this experience a review process was put in place, with the client involved, to discover what went wrong and how this could be averted in future. A report was written with a set of positive recommendations regarding procedures for future projects.

A project does not need to have been a failure before it is deemed necessary to carry out a review; there is much to be learnt from successful projects too. The purpose of a review is to consider how the approach to a project could be improved and in this way to make recommendations for future studies; whenever possible specific actions should be identified. This is primarily a positive role. It also enables the client to give their perspective on the project which is no doubt good for customer relations as well as a useful learning experience for all. On large projects it may be useful to hold reviews at regular intervals during the project in order to monitor progress and put corrective actions in place.

During the review, the following questions should be asked by the project team:

- What was done well?
- What could have been done better?
- How could the approach be improved next time (identifying specific actions)?

The discussion should focus on the process, the sequence of the project phases and their management, and the content, how each phase was performed. Among the subjects for discussion are: the project management procedures, the methods employed in performing each phase (from setting the objectives to implementation), whether the objectives of the project were met, whether the work was completed on time and the level of client satisfaction. Always aim to have discussed at least three things that were done well before any comments are made on what could have been done better; in this way the conversation is more likely to be about positive improvement than negative criticism. Discussion around the first two questions should lead to conclusions regarding the third. At the end of the review a summary of the main findings should be written ready to be implemented in the next simulation project.

13.5 Performing further simulation work

Having completed the simulation project there may be further simulation studies that need to be performed. There are three possible routes that can be taken:

1 *Keeping the model up to date* In some instances it is useful to keep the simulation model up to date, enabling continued analysis of the facility in line with its current status. This is the case, for example, when the model continues to be used throughout implementation. A procedure needs to be put in place to ensure that the modeller is notified of any design changes, that the simulation is changed and that the validity checked. During this process there is also potential for improving the accuracy of the model data, especially unobtainable category C data (Sec. 7.3).

2 *A further phase of the same project* At the beginning of the project further phases of work may already have been identified, for example an initial project which only looked at stage 1 of a facility which is being commissioned in two stages. Sometimes the need for further phases of work are only recognized as the project progresses, for instance the experimental results show an unexpected difficulty. If some time has passed since the model was used, it should not be assumed that the simulation is still valid; the model should be compared to the current status of the facility design. Any changes that are required can be made and the model validated before any further phases of work are performed.

3 *A new project* A fresh problem may need to be tackled using simulation. If previous projects have been successful it is easier to convince clients that simulation should be used again.

Conclusion

The simulation project is complete and hopefully it has been a success, both in terms of the way in which it was performed and through the impact of the results, recommendations and implementation. Many lessons have no doubt been learnt which can now be taken into future simulation studies.

Summary

Communicating the results

- Why communicate the results?: client satisfaction, developing new ideas, obtaining agreement to the recommendations
- Methods of communicating the results: written reports, verbal presentations
- Outline of report/presentation: executive summary (objectives, key results, conclusions and recommendations), introduction to the problem and project objectives, model summary, outline of experimentation, key results, conclusions and recommendations, further work, appendices (detailed results)
- Ensure that the results and conclusions are understood
- Gain agreement for the recommendations

Implementing the recommendations

- Identify who will implement the recommendations: the simulation project team, a new project team, a joint effort
- Develop an implementation plan: the recommendations to be implemented, how the recommendations will be implemented, individual responsibilities, when the implementation (of each phase) will be complete
- Provide an implementation project specification
- Monitor the outcome: differences occur due to faulty implementation, changes to the operating conditions, invalid simulation models

Completing the documentation

- Project specification
- Model documentation
- User documentation
- Final report
- Minutes of project team meetings
- Implementation project specification
- Project review

Reviewing the project

- What was done well?
- What could have been done better?
- How could the approach be improved next time (identifying specific actions)?

Performing further simulation work

- Keeping the model up to date
- A further phase of the same project
- A new project

Case study 13.1: Panorama Televisions

An example report is prepared for Panorama Televisions. Note that an executive summary is not given since the report is in itself a summary.

Introduction to the problem and project objectives
Panorama Televisions have been involved in the manufacture of electrical goods since the early days of the radio. They now concentrate on the production of high-quality, premium priced televisions for the international market. There are four televisions in their product range: small, medium, large and flat screen.

Last year, to meet increased demand, Panorama invested in a new television assembly plant. However, the plant has never achieved its target throughout of 500 units per day; in fact, throughput is only just over 400 units.

The problem has been investigated with a simulation model. The overall aim of this work was:

- To achieve a target throughput of 500 units per day from the television assembly line

The specific objectives were:

- To determine whether 500 units per day can be achieved with additional pallets only
- To identify the additional storage and pallets required to achieve 500 units per day

The model, experimentation, results, conclusions and recommendations are now summarized in this report.

Model summary
The model includes the television assembly plant specifically from OP10 (load television to pallet) to OP60 (unload television from pallet). The final test and pack were not modelled since it was already known that there is enough capacity in this area.

The following details were included in the model:

- Production schedule
- Number of pallets
- Automatic operations: cycle times, breakdowns, set-ups on changeover, repair/set-up labour
- Manual operations: cycle times
- Conveyors: capacity, transfer times
- Re-work: percentage test failures, re-work times

The following were assumed:

- Conveyor breakdowns are infrequent
- Sub-components, such as television boxes and cathode ray tubes are 100 per cent available
- No work, including repairs to broken machines, takes place over a weekend

Therefore, none of these was modelled.

Outline of experimentation
Initial experimentation was performed with additional pallets only and then the effect of increasing buffers was investigated. This identified the need for an additional repair/set-up operator, after which full experimentation was carried out. The experiments and the results are now discussed.

Key results
A summary of the experiments and the results are given in Table 13.1. All the experiments were performed with two repair/set-up operators

Conclusions and recommendations
The following conclusions have been made:

- An additional repair/set-up operator is required
- Using the correct number of pallets and increasing the buffering has a significant effect on throughput
- A throughput of 490 units per day can be achieved with a 200 per cent increase in buffering and 125 pallets
- It is not possible to achieve target throughput by increasing the number of pallets and quantity of buffering (within defined limits) alone

Since the target throughput has not been achieved additional policies should be considered, for example increasing the buffering by more than 200 per cent, improving maintenance to reduce the number of breakdowns

Table 13.1 Panorama Televisions: results of experimentation

Experiment number	Number of pallets	Percentage increase in buffers	Mean daily throughput	Standard deviation
1	50	0	460.10	35.51
2	75	0	447.40	35.55
3	50	100	461.60	33.23
4	75	100	477.97	38.16
5	100	100	478.23	38.23
6	125	100	478.03	35.70
7	150	100	407.10	40.09
8	100	200	487.93	30.88
9	125	200	489.33	31.83
10	150	200	489.33	31.83
11	175	200	488.50	30.58
12	200	200	478.50	36.03

and changing the production schedule to reduce the number of set-ups. It is probable that the latter two policies would reduce the need for additional buffering and 'pallets. However, there is some doubt concerning Panorama's ability to control these factors. This work was outside the scope of this project and should be considered as part of a separate study.

Further work
Alternative methods for increasing throughput should be investigated, particularly:

• Increasing the size of buffers beyond 200 per cent
• Improving maintenance procedures
• Changing the production schedule

These require both simulation work, to investigate their impact, and a study of their feasibility.

Case study 13.2: Natland Bank

An example report is prepared for Natland Bank. Note that an executive summary is not given since the report is in itself a summary.

Introduction to the problem and project objectives

Natland Bank are planning to open a new branch. They were concerned about the facilities required in order to achieve satisfactory service levels, especially with reference to the automatic teller machines. The bank's policy is that customers should not have to wait for more than three minutes to be served. Natland would also like to know whether it is best for customers to wait in individual queues for each service point or in a single queue.

These issues have been investigated with a simulation model. The objectives of this work were:

- To compare the customer service level achieved with one, two and three auto tellers
- To identify which queuing policy gives minimum average waiting times; the choices are a single queue, or individual queues for each auto teller

The model, experimentation, results, conclusions and recommendations are now summarized in this report.

Model summary

The model includes the operation of the auto tellers with the following details included:

- Customers: arrival rates
- Queues: single/multiple queues, queue discipline and jockeying
- Auto tellers: service times

It was assumed that auto teller failures are negligible and so they were not modelled.

Outline of experimentation

The following experiments were performed:

- 1 teller, 1 queue
- 2 tellers, 1 queue
- 2 tellers, 2 queues
- 3 tellers, 1 queue
- 3 tellers, 3 queues

Table 13.2   Natland Bank: results of experimentation

Experiment	Mean queuing time (mins)	Percentage of customers within 3 min	Teller utilization (%)
1 teller, 1 queue	n/a	4.87	n/a
2 tellers, 1 queue	2.26	70.71	80.75
2 tellers, 2 queues	2.30	70.32	80.35
3 tellers, 1 queue	0.12	100.00	54.29
3 tellers, 3 queues	0.35	100.00	53.99

Key results
A summary of the results is given in Table 13.2.

Conclusions and recommendations
The following conclusions have been made:

• One teller is not capable of meeting customer service requirements
• With two tellers 70 per cent of customers are served within three minutes of arriving at a queue, although some customers have to wait for up to 10 minutes at peak times
• With three tellers all customers are served within three minutes of arriving at the auto teller queues
• Teller utilization is on average 80 per cent with two tellers and 55 per cent with three
• A single queue gives a slight improvement in service with three tellers but for two tellers there is no significant difference; this is due to the level of queue jockeying

Two or three auto tellers should be purchased. The improvement in customer service with three auto tellers needs to be weighed against the additional cost. A single queue may give some slight improvement in customer service and should therefore be considered.

Further work
An investigation should be made of the other bank facilities, the manual tellers and the enquiries desk.

PART SIX

Statistical methods in simulation

14

Statistical methods in simulation

The aim of this book has been to provide practical 'rules of thumb' which improve the use of simulation. However, a large number of statistical techniques are available which enable more thorough analysis of both the simulation inputs and the results. It is the purpose of this chapter to describe four of these techniques, namely:

- Testing the fit of distributions: the chi-square test
- Choosing a warm-up period: Welch's method
- Assessing the accuracy of point estimates: confidence intervals
- Comparing alternatives: visual inspection of confidence intervals

At the beginning of the chapter some general terminology is introduced before describing each method in turn. The methods are demonstrated with either the example of Panorama Televisions or Natland Bank. In some instances statistical tables are required and relevant tables can be found in Appendix 3. For those wishing to take a more detailed look at statistical methods that are relevant to simulation, further reading is suggested at the end of the book.

14.1 General terminology
Before describing any statistical methods it is useful to understand some general principles and terminology.

HYPOTHESIS TESTS
A hypothesis is a proposition that is assumed to be true without solid proof. A hypothesis test aims to establish, with a greater degree of certainty, that the hypothesis is true, or otherwise. For instance, a test can be

performed to verify the hypothesis that a standard statistical distribution is a good fit to some data.

Two types of error can be made when a hypothesis test is being performed. A type I error occurs when a correct hypothesis is wrongly rejected and a type II error occurs when a hypothesis that is false is in fact accepted. The level of significance is the probability that a type I error might occur; typical values are 5 and 1 per cent. The lower the level of significance the greater the probability of a type II error. Therefore, the level of significance should be selected to reflect the relative risk of incurring type I or type II errors.

14.2 Testing the fit of distributions: the chi-square test

There are a number of instances in which it is useful to compare the fit of two distributions, for example when fitting statistical distributions to some collected data (the last part of Sec. 7.5) and in validating the results of a model against some historic data (the second part of Sec. 10.2). The chi-square test, in which the hypothesis is that two distributions are equivalent, can be used to perform these analyses. The example in Sec. 7.5, based on the repair times of a machine, is now used to demonstrate this.

Table 14.1 shows the distribution of data that had been collected on repair times. It was proposed (or hypothesized) that these data could be represented by one of three Erlang distributions, each with a mean of 7.6 minutes, and shape parameters of one, three and five respectively. The chi-square test is now performed on these data.

Table 14.1 Example: repair time data

Repair time (mins)	Observed frequency
0.0–3.0	10
3.0–6.0	37
6.0–9.0	23
9.0–12.0	14
12.0–15.0	7
15.0–18.0	5
18.0–21.0	2
> 21.0	2
Total	100

Table 14.2   Example: expected frequencies for Erlang distributions

Repair time (mins)	Expected frequency Erlang (7.6, 1)	Expected frequency Erlang (7.6, 3)	Expected frequency Erlang (7.6, 5)
0.0–3.0	32.8	11.6	5.0
3.0–6.0	21.7	30.6	31.1
6.0–9.0	14.8	26.6	34.2
9.0–12.0	10.1	16.4	19.0
12.0–15.0	6.7	8.5	7.6
15.0–18.0	4.5	3.7	2.3
18.0–21.0	3.1	1.6	0.6
> 21.0	6.3	1.0	0.2
Total	100.0	100.0	100.0

GENERATE THE EXPECTED FREQUENCIES

The actual data must be compared to the expected frequencies from the proposed distribution. Therefore, the first step is to generate these expected frequencies. For some distributions this is straightforward; for example the values from a uniform distribution can simply be calculated. However, for other distributions this is not as simple and other methods need to be employed. A useful approach is to generate the expected frequencies in the simulation software. By taking a large number of samples, say 100 000, from the proposed distribution and recording them in a histogram with the same ranges as the collected data a good approximation can be made. The results for the three Erlang distributions are shown in Table 14.2. Note that the data sum to a hundred in order to make comparison with the data in Table 14.1 possible.

CALCULATE THE CHI-SQUARE VALUE

The chi-square value is calculated as follows:

$$\text{Chi-square value} = \sum_{i=1}^{k} \frac{(O_i - E_i)^2}{E_i}$$

where

$$O_i = \text{observed frequency in } i\text{th range}$$
$$E_i = \text{expected frequency in } i\text{th range}$$
$$k = \text{total number of ranges}$$

Table 14.3 shows the results of this calculation for the repair time example.

Table 14.3 Example: calculation of chi-square value

Repair time (min)	$(O_i - E_i)^2/E_i$ Erlang (7.6, 1)	$(O_i - E_i)^2/E_i$ Erlang (7.6, 3)	$(O_i - E_i)^2/E_i$ Erlang (7.6, 5)
0.0–3.0	15.85	0.22	5.00
3.0–6.0	10.79	1.34	1.12
6.0–9.0	4.54	0.49	3.67
9.0–12.0	1.51	0.35	1.32
12.0–15.0	0.01	0.26	0.05
15.0–18.0	0.06	0.46	3.17
18.0–21.0	0.39	0.10	3.27
> 21.0	2.93	1.00	16.20
Total	36.08	4.22	33.80

SELECT A LEVEL OF SIGNIFICANCE AND DETERMINE THE NUMBER OF
DEGREES OF FREEDOM
Having generated the chi-square value, a level of significance is selected; in
this example 5 per cent is used. A further factor also needs to be considered,
the 'degrees of freedom'. The exact meaning of the degrees of freedom is of
little importance here; however, its value is calculated as follows:

$$\text{Degrees of freedom} = \text{number of ranges} - 1$$

In the repair time example the number of ranges is eight; therefore the
number of degrees of freedom is 7.

PERFORM THE CHI-SQUARE TEST
Finally, the chi-square value is compared to a critical chi-square value read
from the table in Sec. A3.1 in Appendix 3. The critical value is obtained for
the correct level of significance and the number of degrees of freedom. In
the example it is 14.0671. If the chi-squre value is less than the critical value
then the hypothesis is accepted and the proposed distribution cannot be
rejected as a good fit. If the chi-square value is more than the critical value
then the hypothesis and the proposed distribution are rejected.
  The results for the example are shown in Table 14.4. The Erlang distribu-

Table 14.4 Example: results of the chi-square test

Proposed distribution	Chi-square value	Critical value	Accept/reject
Erlang (7.6, 1)	36.08	14.0671	Reject
Erlang (7.6, 3)	4.22	14.0671	Accept
Erlang (7.6, 5)	33.80	14.0671	Reject

tion with a mean of 7.6 minutes and shape parameter of three seems suitable.

Further distributions could now be tested, for example Erlang distributions with shape parameters of two and four. Another distribution may give a better fit which would be shown by a smaller chi-square value.

14.3   Choosing a warm-up period: Welch's method

A practical 'rule of thumb' for choosing the warm-up period is described in Sec. 11.3 (first part). Now an alternative approach, Welch's method, is described and illustrated by Panorama Televisions.

PERFORM A NUMBER OF RUN REPLICATIONS

The model is run a number of times with different pseudo-random number streams. Between five and ten replications are recommended, and more if the model behaves in a particularly random fashion. The length of each run should be much longer than the anticipated warm-up period. A time-series of the key result is collected for every replication and the mean average for each period calculated.

In the example Panorama Televisions five replications have been performed, each of 48 hours' duration, and results on hourly throughput were collected. The mean hourly throughput for each hour, across all five replications, was then calculated (Table 14.5).

CALCULATE MOVING AVERAGES

Moving averages are then calculated for the key result. 'Windows' of various sizes are selected and the moving averages calculated as follows:

$$
\text{Moving average}_i =
\begin{cases}
\dfrac{\displaystyle\sum_{s=(i-1)}^{i-1} \text{mean}_{i+s}}{2i-1} & \text{if } i = 1, \ldots, w \\[3em]
\dfrac{\displaystyle\sum_{s=-w}^{w} \text{mean}_{i+s}}{2w+1} & \text{if } i = w+1, \ldots, m-w
\end{cases}
$$

where

i = period for which the moving average is being calculated
mean = mean of the key result, across all the replications, for period i
w = number of periods in the window
m = number of periods in the run

This is best illustrated with an example. For Panorama Televisions a window of five periods has been selected and applied to the data in Table

14.5. For period 8, for instance, the moving average was calculated from the results for the five periods preceding period 8, period 8 itself and the five following periods:

Moving average$_8$

$$= \frac{\text{hour 3 throughput} + \cdots + \text{hour 8 throughput} + \cdots + \text{hour 13 throughput}}{11}$$

Moving average$_8$

$$= \frac{17.40 + \cdots + 12.20 + \cdots + 20.20}{11} = 16.87$$

However, for period 2, since the full window is not available, the moving average was calculated from the results for periods 1, 2 and 3 only:

Moving average$_3$

$$= \frac{\text{hour 1 throughput} + \text{hour 2 throughput} + \text{hour 3 throughput}}{3}$$

Moving average$_3$

$$= \frac{3.60 + 18.20 + 17.40}{3} = 13.07$$

The moving averages are calculated for various windows. For obvious reasons the number of periods in the window cannot be more than half the periods in the run. In Panorama Televisions moving averages have been calculated for windows of 5, 10 and 20, as shown in Table 14.5.

DETERMINE THE WARM-UP PERIOD
Time-series are then constructed for each moving average. After a number of periods the graph should become relatively flat if the model has reached steady-state. The warm-up period is selected from the point at which the graph becomes flat. If the graph does not flatten out then longer windows should be examined until a smooth line is obtained. However, the window should be kept as small as possible and the warm-up period selected from the smallest window for which a flat line is obtained.

The moving averages for Panorama Televisions are shown in Fig. 14.1. Although a window of 10 gives a reasonably flat line, an improvement is

Table 14.5 Panorama Televisions: mean hourly throughput and calculation of moving averages

Hour	Mean hourly throughput (5 replications)	Moving average window = 5	Moving average window = 10	Moving average window = 20
1	3.60	3.60	3.60	3.60
2	18.20	13.07	13.07	13.07
3	17.40	14.56	14.56	14.56
4	20.60	14.06	14.06	14.06
5	13.00	14.73	14.73	14.73
6	17.80	15.47	15.47	15.47
7	7.80	16.69	15.95	15.95
8	12.20	16.87	16.11	16.11
9	22.00	16.85	15.80	15.80
10	18.60	16.53	16.06	16.06
11	19.00	16.42	16.14	16.14
12	17.00	16.18	16.83	16.60
13	20.20	17.36	17.14	16.88
14	17.20	17.69	17.30	16.81
15	17.00	17.25	17.25	16.67
16	11.80	17.07	17.41	16.74
17	15.20	16.98	17.30	16.65
18	20.80	17.69	17.57	16.62
19	15.80	17.75	17.75	16.68
20	17.20	17.95	17.74	16.74
21	16.60	17.89	17.50	16.75
22	18.00	18.24	17.24	17.11
23	24.80	18.07	17.26	17.00
24	20.80	17.64	17.06	17.03
25	19.40	18.18	17.01	16.92
26	16.40	17.85	17.12	16.77
27	15.60	17.56	17.31	16.85
28	13.40	17.51	17.29	17.16
29	16.00	16.71	17.30	
30	21.80	16.29	17.66	
31	13.60	16.29	17.35	
32	13.40	16.24	17.44	
33	17.40	16.15	17.23	
34	16.00	16.85	16.93	
35	16.20	17.51	16.70	
36	19.40	16.51	16.10	
37	15.80	16.95	16.34	
38	14.60	16.96	16.56	
39	21.20	17.07		
40	23.20	17.07		
41	10.80	16.22		
42	18.40	16.40		
43	13.60	16.80		
44	18.60			
45	16.00			
46	6.80			
47	21.40			
48	20.20			

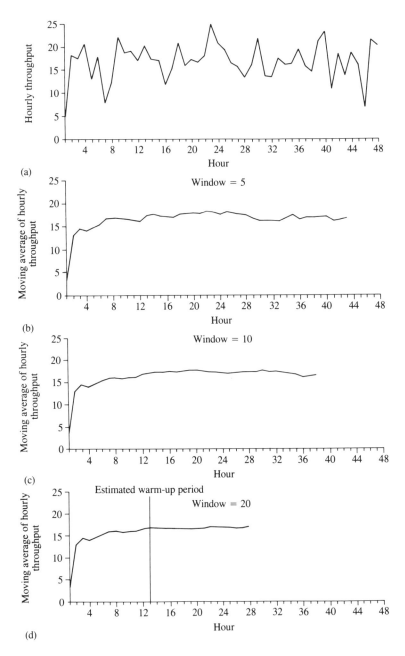

Fig. 14.1 Panorama Televisions: mean hourly throughput and moving averages across five replications

obtained with a window of 20. Therefore, a warm-up period of 13 hours has been selected from that time series.

14.4   Assessing the accuracy of point estimates: confidence intervals

During results analysis point estimates are used to determine the success of an experiment, for example estimates of the average daily throughput or mean waiting time (Sec. 12.2). However, these are only estimates since the simulation experiments produce just a sample of the potential results. Confidence intervals are a means for expressing the degree of accuracy of the point estimate. It consists of an upper and lower limit within which the true value of the result is expected to lie; the probability of this is specified by the level of significance. For a given level of significance, if the length of the interval is small then high accuracy has been achieved; if it is large then the accuracy is lower. Some simulation packages automatically report confidence intervals. This section describes how a confidence interval is calculated and illustrates its use with the case study of the Natland Bank.

CALCULATING A CONFIDENCE INTERVAL

Firstly, a number of replications are performed and then point estimates of the key results are calculated. In the Natland Bank case study (Chapter 11), 20 replications are performed. In order to demonstrate how confidence intervals are calculated the results of the first three replications will be taken, namely mean queuing times of 2.27, 1.57 and 1.49 minutes respectively. By calculating the mean of these results a point estimate of 1.78 minutes is obtained.

Secondly, the standard deviation of these results is calculated using the formula:

$$\text{Standard deviation} = \sqrt{\frac{\sum_{i=1}^{n} (\text{result}_i - \text{mean})^2}{n - 1}}$$

where n = number of replications. Therefore in the case study of the Natland Bank:

$$\text{Standard deviation} = \sqrt{\frac{(2.27 - 1.78)^2 + (1.57 - 1.78)^2 + (1.49 - 1.78)^2}{3 - 1}}$$

$$= 0.43$$

A level of significance is selected, say 5 per cent, giving a 95 per cent chance that the interval contains the true value of the result. Student's t-distribution (Sec. A3.2 of Appendix 3) is used to obtain a value t based on the significance level and the number of replications. Since the confidence interval provides

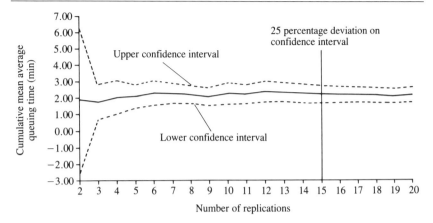

Fig. 14.2 Natland Bank: confidence interval on cumulative mean average queuing time to determine the number of replications

both a lower and upper limit the significance level is divided by two in order to split the probability between the two extremes; for a 5 per cent significance level, 2.5 per cent significance is selected from Student's *t*-distribution. The confidence interval is then calculated as follows:

$$\text{Confidence interval} = \text{mean} \pm t_{n-1,\,\alpha/2}\, \frac{\text{standard deviation}}{\sqrt{n}}$$

where

$$n = \text{number of replications}$$
$$t_{n-1,\,\alpha/2} = \text{value from Student's } t\text{-distribution for } n - 1 \text{ and the}$$
$$\text{significance level } \alpha/2$$

In the Natland Bank case study:

$$\text{Confidence interval} = 1.78 \pm 4.303\, \frac{0.43}{\sqrt{3}}$$

Lower limit	= 0.71
Upper limit	= 2.84

Therefore, there is a 95 per cent probability that the true value of the mean queuing time is in the interval 0.71 to 2.84 minutes.

USING CONFIDENCE INTERVALS TO DETERMINE THE NUMBER OF REPLICATIONS

Confidence intervals can be used to determine the number of replications required (first part of Sec. 11.4). Replications are continued until a

Table 14.6 Natland Bank: confidence interval on cumulative mean average queuing time to determine the number of replications

Replication	Mean queuing time (min)	Cumulative mean average queuing time (min)	Standard deviation	95% confidence interval Lower confidence limit	95% confidence interval Upper confidence limit	Percentage deviation of limit from mean
1	2.27	2.27	n/a	n/a	n/a	n/a
2	1.57	1.92	0.49	− 2.53	6.37	231.62
3	1.49	1.78	0.43	0.71	2.84	60.00
4	2.86	2.05	0.65	1.02	3.07	50.13
5	2.29	2.10	0.57	1.39	2.80	33.71
6	3.30	2.30	0.71	1.55	3.04	32.34
7	2.18	2.28	0.65	1.68	2.88	26.26
8	1.64	2.20	0.64	1.66	2.74	24.35
9	1.02	2.07	0.72	1.52	2.62	26.63
10	3.92	2.25	0.89	1.61	2.89	28.38
11	1.79	2.21	0.86	1.63	2.79	26.11
12	4.07	2.37	0.98	1.74	2.99	26.30
13	1.92	2.33	0.95	1.76	2.90	24.51
14	1.01	2.24	0.98	1.67	2.80	25.16
15	1.68	2.20	0.95	1.67	2.73	23.93
16	1.92	2.18	0.92	1.69	2.67	22.48
17	2.09	2.18	0.89	1.72	2.64	21.07
18	1.22	2.12	0.89	1.68	2.57	20.94
19	1.24	2.08	0.89	1.65	2.51	20.71
20	3.98	2.17	0.97	1.72	2.63	20.83

confidence interval of a certain accuracy is achieved, say within ± 10 per cent of the mean. This is demonstrated by the 20 replications performed for the Natland Bank in Chapter 11. Confidence intervals have been calculated for each additional replication and shown in Table 14.6 and graphically in Fig. 14.2.

Note how the confidence interval narrows with further replications. The point at which the interval gives a deviation of 25 per cent from the mean, 15 replications, has been noted and could be used as an estimate of the number of replications required. For greater accuracy more replications would be necessary.

APPLICABILITY OF CONFIDENCE INTERVALS

The basic assumption behind the calculation of a confidence interval is that the data are independent. Therefore, a confidence interval can only be calculated for independent replications and not for results from a single long run. More complex methods are available which enable confidence intervals to be constructed for long simulation runs, but these are not discussed here.

14.5   Comparing alternatives: visual inspection of confidence intervals

In Sec. 12.2 (second part) a method for comparing point estimates is discussed. This procedure can be improved by comparing confidence intervals for each of the point estimates. For example, in order to compare four different options run replications could be performed and then confidence intervals calculated for each of them, as represented in Fig. 14.3.

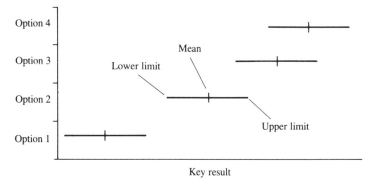

Fig. 14.3   Comparing confidence intervals for four different options

Option 2 is significantly different from option 1 since the two confidence intervals do not overlap. Option 3 is probably significantly different from option 2; however, there must be some caution since the intervals overlap. Option 4 cannot be accepted as significantly different from option 3 since the mean result of option 4 falls within the confidence interval for option 3. If it is important to improve this information, further run replications could be performed which would increase the accuracy of the point estimates and narrow the confidence intervals.

Such an analysis could be performed on the mean queuing time results for the Natland Bank (Chapter 12, Tables 12.8 to 12.11). Confidence intervals have been calculated in Table 14.7 and shown visually in Fig. 14.4. Note that the number of replications is 6 and $t_{5, 2.5\%} = 2.571$.

The results show that there is a significant difference in queuing time between two and three tellers. With two tellers there is no difference between one and two queues. However, with three tellers a single queue gives a significant improvement over three separate queues.

Conclusion

This chapter has demonstrated four useful statistical methods. These enable

Table 14.7   Natland Bank: confidence intervals for mean queuing time

Number of tellers	Number of queues	Mean queuing time (min)	Standard deviation	95% confidence interval	
				Lower limit	Upper limit
2	1	2.26	0.67	1.56	2.96
2	2	2.30	0.71	1.55	3.05
3	1	0.12	0.02	0.10	0.14
3	3	0.35	0.03	0.32	0.38

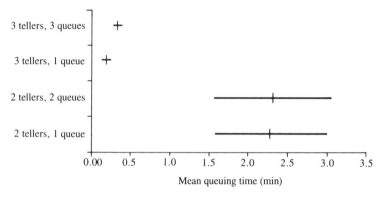

Fig. 14.4   Natland Bank: confidence intervals for mean queuing time

a more thorough analysis of the simulation inputs and the results obtained. However, many more methods are available and these can be found in the further reading suggested at the end of the book.

Summary
General terminology

- Hypothesis tests
- Level of significance

Testing the fit of distributions: the Chi-square test

- Generate the expected frequencies
- Calculate the chi-square value:

$$\text{Chi-square value} = \sum_{i=1}^{k} \frac{(O_i - E_i)^2}{E_i}$$

- Select a level of significance and the number of degrees of freedom
- Perform the chi-square test: if the chi-square value is less than the value obtained from the chi-square table, then the distribution cannot be rejected

Choosing a warm-up period: Welch's method

- Perform five to ten run replications, more if necessary
- Calculate moving averages based on various windows (w):

$$
\text{Moving average}_i = \begin{cases} \dfrac{\displaystyle\sum_{s=(i-1)}^{i-1} \text{mean}_{i+s}}{2i-1} & \text{if } i = 1, \ldots, w \\[4mm] \dfrac{\displaystyle\sum_{s=-w}^{w} \text{mean}_{i+s}}{2w+1} & \text{if } i = w+1, \ldots, m-w \end{cases}
$$

- Determine the warm-up period: draw time-series of moving averages, select a smooth time-series with the shortest window; the warm-up period is the point at which the line becomes flat

Assessing the accuracy of point estimates: confidence intervals

- Calculating confidence intervals:

$$
\text{Confidence interval} = \text{mean} \pm t_{n-1,\,\alpha/2} \frac{\text{standard deviation}}{\sqrt{n}}
$$

- Using confidence intervals to determine the number of replications: perform replications until the point estimate achieves a certain accuracy
- Confidence intervals can only be calculated for results from multiple replications and not for long runs

Comparing alternatives: visual inspection of confidence intervals

- If the confidence intervals do not overlap there is a significant difference
- If the confidence intervals overlap but the mean of one option is outside the interval of another, a significant difference can be accepted with some caution
- If the confidence intervals overlap and the mean of one option is inside the interval of another, a significant difference cannot be accepted

Epilogue

Since 1986 and my first simulation 'success' I have worked on a large number of simulation projects, either performing the work myself or acting as an advisor. There have been many rewarding moments: projects that have been completed on time, results that have proved beneficial and clients who have been satisfied and expressed their gratitude. There have, of course, been the not-so-rewarding moments: staying up late in an attempt to claw back time, experiments that have given little additional insight and clients who have, rightly or wrongly, complained. Through these experiences I have discovered much and the aim of this book has been to summarize that which has been learnt. By now the reader should:

- Know the stages required for a simulation project
- Be able to formulate a problem for solution by simulation
- Know how to build valid and credible models
- Be able to perform simulation experiments, analyse the results and draw conclusions
- Know how to successfully manage a simulation project

Throughout, the purpose has been to show that simulation is not only a science but also an art. The individual steps that have been discussed should be performed in an iterative manner, moving from problem definition to project completion and implementation. The focus has been on practical methods and 'rules of thumb' through which the use of simulation can be improved. It would be dangerous to claim that all of the methods and procedures should be followed exactly in every project; they certainly need to be adapted for an individual circumstance.

The benefits of simulation are clear: risk reduction, greater understanding, operating cost reduction, lead time reduction, faster plant changes, capital cost reduction and improved customer service. There are also many software packages available, both simulation languages and simulators. However, the successful use of simulation remains relatively rare. This is most likely caused by low awareness and by a lack of success in performing simulation projects. Awareness necessarily takes time to improve while successful projects require skills in both simulation and project management—the skills that have been discussed in this book.

Appendices

APPENDIX 1

Simulation societies and journals

The following societies and journals are a useful source of information on simulation and simulation software:

The Society for Computer Simulation
PO Box 17900
San Diego
California CA 92177
USA

Main journal: *Simulation*

Institute of Industrial Engineers
25 Technology Park / Atlanta
Norcross
Georgia GA 30092-2988
USA

Main journal: *Industrial Engineering*

Also, any national Operational Research societies, for example:

Operational Research Society
Neville House
Waterloo Street
Birmingham B2 5TX
UK

Main journals: *Journal of the Operational Research Society*, *OR Insight*

Operations Research Society of America
1314 Guilford Avenue
Baltimore
Maryland MD 21202
USA

Main journals: *Operations Research, OR/MS Today*

Panorama Televisions and Natland Bank model data

A2.1 Panorama Televisions
All times are in minutes

General data
   Production schedule:
      Small TV   20
      Medium TV   30
      Large TV   40
      Flat screen TV   20
   Number of pallets   30

OP10
   Cycle time   Normal, mean = 1.9,
                                   standard deviation = 0.19

OP20
   Cycle time   2.1

   Breakdowns:
      Time between failure   Negative exponential, mean = 300
      Repair time   Triangular, minimum = 5, mode = 25,
                                   maximum = 60
      Number of repair labour   1
   Set-ups on changeover:
      Set-up time   Normal, mean = 5.0,
                                   standard deviation = 0.5

OP30

Cycle time	2.0
Breakdowns:	
Time between failure	Negative exponential, mean = 450
Repair time	Erlang, mean = 35, $K = 3$
Number of repair labour	1
Set-ups on changeover:	
St-up time	Normal, mean = 5.0, standard deviation = 0.5

OP40

Cycle time	2.0 per station 5 stations

Test

Cycle time	1.5
Breakdowns:	
Time between failure	Negative exponential, mean = 250
Repair time	Since the data gives a bimodal distribution, a user distribution is used:

Repair time	Frequency
0.0–10.0	10
10.0–20.0	25
20.0–30.0	20
30.0–40.0	7
40.0–50.0	5
50.0–60.0	17
60.0–70.0	14

Number of repair labour	1
Set-ups on changeover:	
Set-up time	Normal, mean = 3.0, standard deviation = 0.3

OP50

Cycle time	2.1
Breakdowns:	
Time between failure	Negative exponential, mean = 370
Repair time	Triangular, minimum = 10, mode = 30 maximum = 80
Number of repair labour	1
Set-ups on changeover:	
Set-up time	Normal, mean = 5.0, standard deviation = 0.5

OP60
Cycle time   Normal, mean = 1.9,
                                    standard deviation = 0.19

Main line conveyors
Capacity   5
Transfer times   2.5
Type   Free flow roller

Pallet return conveyor
Capacity   40
Transfer times   20.0
Type   Free flow roller

Re-work
Percentage test failures   5.0%
Re-work times   Negative exponential, mean = 35

Labour
Total number of repair/set-up
labour   1

Validation data
Mean daily throughput   407
Distribution of daily
throughput (Fig. A2.1)

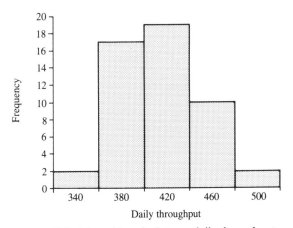

Fig. A2.1   Panorama Televisions: historic data on daily throughput over 50 days

A2.2   Natland Bank

Customers

Arrival rate at auto tellers:

Time of day	Customers/hour
09.30–10.30	95
10.30–11.30	125
11.30–12.30	150
12.30–13.30	190
13.30–14.30	145
14.30–15.30	175

Queue discipline: Customers go to shortest queue and jockey if there are at least three fewer people in an alternative queue

Auto tellers

Service times modelled as a real discrete user distribution:

Transaction type	Service time (s)	Percentage of customers
C	30	40
B	25	10
R	20	8
C, B	50	25
C, R	45	10
B, R	40	2
C, B, R	65	5

C = cash
B = balance of account
R = cheque book request

APPENDIX 3

Statistical tables

This appendix summarizes the statistical tables for the chi-square and Student's t-distributions. More detailed tables are provided in the statistical literature, for example:

Murdoch, J. and Barnes, J. (1986), *Statistical Tables for Science, Engineering, Management and Business Studies*, 3rd ed., Macmillan Education, London.

A3.1 Chi-square distribution

	Level of significance	
Number of degrees of freedom	5%	1%
1	3.84146	6.63490
2	5.99147	9.21034
3	7.81473	11.3449
4	9.48773	13.2767
5	11.0705	15.0863
6	12.5916	16.8119
7	14.0671	18.4753
8	15.5073	20.0902
9	16.9190	21.6660
10	18.3070	23.2093
11	19.6751	24.7250
12	21.0261	26.2170
13	22.3621	27.6883
14	23.6848	29.1413
15	24.9958	30.5779
16	26.2962	31.9999
17	27.5871	33.4087
18	28.8693	34.8053
19	30.1435	36.1908
20	31.4104	37.5662
21	32.6705	38.9321
22	33.9244	40.2894
23	35.1725	41.6384
24	36.4151	42.9798
25	37.6525	44.3141
26	38.8852	45.6417
27	40.1133	46.9630
28	41.3372	48.2782
29	42.5569	49.5879
30	43.7729	50.8922
40	55.7585	63.6907
50	67.5048	76.1539
60	79.0819	88.3794
70	90.5312	100.425
80	101.879	112.329
90	113.145	124.116
100	124.342	135.807

A3.2   Student's *t*-distribution

Number of degrees of freedom	Level of significance	
	2.5%	*0.5%*
1	12.706	63.657
2	4.303	9.925
3	3.182	5.841
4	2.776	4.604
5	2.571	4.032
6	2.447	3.707
7	2.365	3.499
8	2.306	3.355
9	2.262	3.250
10	2.228	3.169
11	2.201	3.106
12	2.179	3.055
13	2.160	3.012
14	2.145	2.977
15	2.131	2.947
16	2.120	2.921
17	2.110	2.898
18	2.101	2.878
19	2.093	2.861
20	2.086	2.845
21	2.080	2.831
22	2.074	2.819
23	2.069	2.807
24	2.064	2.797
25	2.060	2.787
26	2.056	2.779
27	2.052	2.771
28	2.048	2.763
29	2.045	2.756
30	2.042	2.750
40	2.021	2.704
60	2.000	2.660
120	1.980	2.617

Further reading

General texts
The following books give good general coverage of simulation.

Banks, J. and Carson, J. (1984), *Discrete-Event System Simulation*, Prentice-Hall, Englewood Cliffs, N.J.

Fishman, G. (1978), *Principles of Discrete Event Simulation*, John Wiley, New York.

Gogg, T. and Mott, J. (1992), *Improve Quality and Productivity with Simulation*, JMI Consulting Group, Palos Verdes Pnsl., Calif.

Law, A. and Kelton, W. (1991), *Simulation Modeling and Analysis*, 2nd ed., McGraw-Hill International Editions, New York.

Pidd, M. (1992), *Computer Simulation in Management Science*, 3rd ed., John Wiley and Sons, Chichester.

Tocher, K. (1963), *The Art of Simulation*, English Universities Press, London.

Specific subjects
The following books and papers provide useful reading on specific subjects relating to the content of this book. It is by no means an exhaustive list but should act as a useful starting point.

THE BENEFITS OF SIMULATION
Hollocks, B. (1992), A well kept secret?, *OR Insight*, **5** (4), 12–17.

Simulation Study Group (1991), *Simulation in UK Manufacturing Industry*, R. Horrocks (ed.), DTI Study Report, The Management Consulting Group, University of Warwick.

SELECTING SIMULATION SOFTWARE
Holder, K. (1990), Selecting simulation software, *OR Insight*, **3** (4), 19–24.

Law, A. and Haider, S. (1989), Selecting simulation software for manufacturing applications: practical guidelines and software survey, *Industrial Engineering*, **21** (5), 33–46.

Law, A. and McComas, M. (1992), How to select simulation software for manufacturing applications, *Industrial Engineering*, **24** (7), 29–35.

Pidd, M. (1989), Choosing discrete simulation software, *OR Insight*, **2** (3), 22–23.

Van Breedam, A., Raes, J. and Van de Velde, K. (1990), Segmenting the simulation software market, *OR Insight*, **3** (2), 9–13.

JUSTIFYING SIMULATION
Mott, J. and Tumay, K. (1992), Developing a strategy for justifying simulation, *Industrial Engineering*, **24** (7), 38–42.

OVERVIEW OF SIMULATION PROJECTS
Dietz, M. (1992), Outline of a successful simulation project, *Industrial Engineering*, **24** (11), 50–53.

Keller, L., Harrell, C. and Leavy, J. (1991), The three reasons why simulation fails, *Industrial Engineering*, **23** (4), 27–31.

Law, A. and McComas, M. (1989), Pitfalls to avoid in the simulation of manufacturing systems, *Industrial Engineering*, **21** (5), 28–31.

Law, A. and McComas, M. (1990), Secrets of successful simulation studies, *Industrial Engineering*, **22** (5), 47–72.

Olgen, O. (1991), Proper management techniques are keys to a successful simulation project, *Industrial Engineering*, **23** (8), 37–41.

PROJECT SPECIFICATIONS
Ashouri, F. (1993), Making contracts, *OR Insight*, **6** (2), 29–31.

VERIFICATION AND VALIDATION
Baxter, L. and Johnson, M. (1993), Don't implement before you validate, *Industrial Engineering*, **25** (2), 60–62.

Carson, J. (1986), Convincing users of model's validity is challenging aspect of modeler's job, *Industrial Engineering*, **18** (6), 74–85.

Sargent, R. (1982), Verification and validation of simulation models, in *Progress in Modelling and Simulation*, F. Cellier (ed.), Academic Press, London.

Sargent, R. (1988), A tutorial on validation and verification of simulation models, in *Proceedings of the 1988 Winter Simulation Conference*, pp. 33–39.

VARIANCE REDUCTION
James, B. (1985), Variance reduction techniques, *Journal of the Operational Research Society*, **36** (6), 525–530.

STATISTICAL METHODS IN SIMULATION
Kleijnen, J. (1987), *Statistical Tools for Simulation Practitioners*, Marcel Dekker, New York.

Kleijnen, J. and Van Groenendaal, W. (1992), *Simulation, A Statistical Perspective*, John Wiley and Sons, Chichester.

Lewis, P. and Orav, E. (1986), *Simulation Methodology for Statisticians, Operations Analysts and Engineers*, Volume 1, Wadsworth and Brooks, Pacific Grove, Calif.

Index